AF598039

Methods in Molecular Biology

For further volumes:
http://www.springer.com/series/7651

Macro-Glycoligands

Methods and Protocols

Edited by

Xue-Long Sun

Cleveland State University, Cleveland, OH, USA

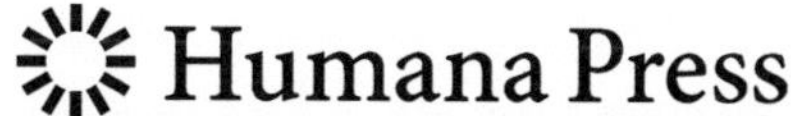

Editor
Xue-Long Sun
Cleveland State University
Cleveland, OH, USA

ISSN 1064-3745 ISSN 1940-6029 (electronic)
Methods in Molecular Biology
ISBN 978-1-4939-3129-3 ISBN 978-1-4939-3130-9 (eBook)
DOI 10.1007/978-1-4939-3130-9

Library of Congress Control Number: 2015951795

Springer New York Heidelberg Dordrecht London

Printed on acid-free paper

Humana Press is a brand of Springer
Springer Science+Business Media LLC New York is part of Springer Science+Business Media (www.springer.com)

Preface

Carbohydrate recognition is a crucial event in many biological processes, such as cell-cell signaling, immune recognition events, pathogen/host interactions, tumor metastasis, tissue growth and repair, etc. Therefore, carbohydrate recognition has come to the forefront of biological scientific research aiming to uncover the molecular mechanisms of many physiological and pathological processes and discover potential therapeutic targets or diagnostic mechanisms for various diseases. Cell surface carbohydrates, existing as glycoproteins, glycolipids, or proteoglycans, are often involved in these biological processes. Consequently, reconstitution of cell surface carbohydrate epitopes to mimic authentic compositions and presentations has become the key research in either studying carbohydrate recognition or developing therapeutic and diagnostic tools.

It has been known that multivalent interactions facilitate both specificity and affinity in carbohydrate–protein interactions, referred to as the "cluster glycosidic effect." In the past decades, glycopolymers, namely, polymers with carbohydrate pendant groups, have been extensively explored as multivalent carbohydrate ligands for studying on carbohydrate–protein interactions and for important biomedical applications. For example, glycopolymers can act as agonists or antagonists for understanding the molecular mechanisms of many biological processes and also provide tremendous opportunities for therapeutic applications. In addition, glycopolymers can serve as potential receptors for biochip/biosensor development, which can be used for understanding carbohydrate–protein interaction, substrate specificity of carbohydrate-processing enzymes, antibody profiling, biomarkers, and pathogen and toxin identification applications. Therefore, there is a high demand for developing facile methods and protocols for synthesizing and characterizing glycopolymers of different interests. This book aims to give the reader detailed research methods and protocols for the synthesis, characterization, and biomedical applications of glycopolymer-based macro-glycoligands.

In this book *Macro-Glycoligands: Methods and Protocols*, we have received excellent contributions from experts in the field. Altogether 17 book chapters cover the recent advances in carbohydrate chemistry and polymer chemistry, and glycobiology aimed at understanding and controlling the outcomes of carbohydrate recognition with particular emphasis on glycopolymer-based macro-glycoligand. The book content is divided into three parts: (I) Synthesis and characterization of glycopolymers; (II) Glycopolymer-nanoparticle conjugates; (III) Surface immobilized glycopolymers. The compilation of these book chapters provides a comprehensive and practical methods and protocols and timely reference to the state of the art in recent glycopolymer research and applications.

There are seven chapters (Chapters 1–7) in *Part I*, which covers recent advances in the synthesis and characterization of glycopolymers and their biomedical applications as well. Glycopolymers carrying pendant sugar moieties can be synthesized by either direct polymerization of carbohydrate-containing monomers with protection group or without protection group, by the postpolymerization conjugation of glycans and synthetic polymers, or by grafting of carbohydrate ligand to polymeric materials. Over the past decades, a variety of direct polymerization methods have been developed, including cyanoxyl-mediated free-radical polymerization (CMFRP), reversible addition-fragmentation chain transfer (RAFT) polymerization, ring opening metathesis polymerization (ROMP), cationic ring-opening polymerization (CROP), ring-opening polymerization (CROP), and atom transfer radical polymerization (ATRP). Chapter 1 provides a straightforward synthesis of chain-end functionalized glycopolymers via CMFRP, in which no protection/deprotection is needed. In Chapter 2, a protecting-free synthesis of well-defined glycopolymers via RAFT is fully demonstrated. Chapter 3 provides a unique method for generating end-labeled amino terminated monotelechelic glycopolymers by ROMP. Chapter 4 provides a facile protecting-group-free synthetic approach to glycopolymers bearing large biologically relevant oligosaccharides having sialic acids via reversible addition-fragmentation chain transfer polymerization (RAFCTP). Chapter 5 presents a detailed methodology to functionalize poly(2-ethyl-2-oxazoline) in a stereoselective manner with a range of carbohydrates that can serve as biological targeting units. Chapter 6 presents a method for the in situ functionalization and (co-) polymerization of allylglycine *N*-carboxyanhydride in a facile one-pot procedure, combining radical thiol-ene photochemistry and nucleophilic ring-opening polymerization techniques, to yield well-defined heterofunctional glycopolypeptides. Finally, Chapter 7 describes a method for preparation of graft copolymers with glycosaminoglycan side chains, which mimic the structure and composition of proteoglycans.

Part II (Chapters 8–14) covers recent advances in the synthesis and characterization of glycopolymer-nanoparticle conjugates and their biomedical applications as well. Nanomaterials are a promising tools for biomedical research and applications, as it is predicted to be beneficial in tackling clinical problems. Glycopolymer-based nanostructures are invaluable tools to both study biological phenomena and design future targeted drug delivery systems. On the other hand glyconanoparticles (GNPs), such as sugar-coated gold, iron oxide, or semiconductor nanoparticles, have magnetic or fluorescence properties, making this multivalent glyco-scaffold suitable for carrying out studies on carbohydrate-mediated interactions and applications in molecular imaging and targeted drug delivery applications. In this book, Chapter 8 describes the methods to prepare well-defined glycopolymer-containing diblock copolymers via RAFT, to self-assemble these macromolecules and to start assessing the in vitro interactions of the self-assembled structures with live cells. Recently, high luminescence, single excitation narrow emission, low photobleaching properties and low toxicity of high quality water-soluble Quantum Dots (QDs) have attracted attention for in vivo labeling/imaging of cells. Chapter 9 describes a synthetic approach to biotinylated glycopolymer functionalized quantum dots, with special emphasis on the development of high quality water-soluble and bioactive QDs with low toxicity for fluorescent probes in biomedical applications. Both ATRP and RAFT polymerization allow for facile surface modification and eventual surface polymerization of monomers exhibiting biological mimicking capabilities. In Chapter 10, ATRP was carried out in a "grafting from" approach to obtain well-defined polypentafluorostyrene modified polymeric microspheres. These fluorinated materials were then converted to fluorinated glycopolymers using a thioglucose salt and thiol-halogen "click" chemistry without the need for any further deprotection chemistry. Further, Chapter 11 describes the synthesis of glycopolymer-grafted polymer particles using two types of surface-initiated living radical polymerization: the first is ATRP, and the other is photoiniferter polymerization. In Chapter 12, a synthetic approach to prepare nonspherical glycopolymer-coated iron oxide nanoparticles is provided, by combining the convenience of inorganic shape control, catecholic chemistry, and thiol-ene reaction. On the other hand, Chapter 13 describes a synthetic approach to glycopolymer-gold(I) nanoparticle conjugate for cancer therapy via three-step strategies to incorporate thiol and dithiocarbamate functionality for the stabilization of gold nanoparticles and the cancer drug for therapeutic application via RAFT polymerization method. Finally, in Chapter 14, a polymer-stabilized glycosylated gold nanoparticle platform was demonstrated with precisely engineered heterotelechelic poly *N*-hydroxyethyl acrylamide polymers bearing a carbohydrate moiety at one end for lectin interaction and a thiol at the other for gold particle attachment.

Part III (Chapters 15–17) covers recent advances in surface immobilization of glycopolymers and their biochip/biosensor development and applications. The presentation of carbohydrates on an array can provide a means to model (mimic) oligosaccharides found on cell surfaces. Tuning the structural features of such carbohydrate arrays can therefore be used to help elucidate the molecular mechanisms of protein-carbohydrate recognition on cell surfaces. Chapter 15 presents a strategy to directly correlate the molecular and structural features of ligands presented on a surface with the kinetics and affinity of carbohydrate–lectin binding. Both Surface Plasmon Resonance (SPR) spectroscopy and atomic force microscopy (AFM) confirmed the spatial distribution of carbohydrate ligands within the surface grafted polymer layer and their lectin binding features. In Chapter 16, a chemoenzymatic synthesis of *O*-cyanate chain-end functionalized sialyllactose-containing glycopolymers and their oriented sialyloligo-macroligand formation for glycoarray and glyco-biosensor applications are demonstrated in detail. This oriented sialyloligo-macroligand platforms are expected to facilitate both affinity and specificity of protein binding and thus provide a versatile tool for profiling glycan recognition via glycoarray and SPR-based glyco-biosensor. The cellular glycocalyx controls many of the crucial signaling pathways involved in cellular development. Synthetic materials that can mimic the multivalency and three-dimensional architecture of native glycans serve as important tools for deciphering and exploiting the roles of these glycans. Chapter 17 describes an approach for remodeling cell surface glycocalyx with glycopolymer-based proteoglycan mimetics that binds FGF2 as a cell-surface engineering strategy to influence stem cell specification.

In this book, we provide a detailed methods and protocols for the synthesis and characterization of glycopolymers and their biomedical applications. Various controlled radical polymerization techniques have been successfully employed for the synthesis of chain-end functionalized glycopolymers with narrow polydispersity. The two significant features of the chain-end functionalized glycopolymers are multivalency, which can help increase the affinity and specificity of bimolecular recognition, and chain-end functional group, which can facilitate direct one-to-one attachment and oriented immobilization of glycopolymers onto solid surfaces for mimicking cell surface carbohydrates. These chain-end functionalized glycopolymers were covalently or noncovalently attached to proteins, nanoparticles, and glass slides in a site-specific fashion and lead to oriented glycopolymer presentation that will find important biomedical applications. Particularly, oriented glycopolymer-based glycan microarrays have exhibited high potential as a high-throughput analytical tool for investigating biological processes engaged with carbohydrates.

As an editor of this *Methods in Molecular Biology Series*, I am very grateful to the *Series Editor John M. Walker* for this opportunity, and I am greatly indebted to all authors, who responded with great enthusiasm to my initial proposal by contributing manuscripts. Also, I would extend my gratitude to Springer for support of this special issue. With respect to the readers, I hope that this compilation of chapters will provide not only practical methods and protocols but also a timely overview and reference to Carbohydrate Recognition and

Application. Further, it will stimulate new ideas for hypothesis-driven research in this certainly fascinating area of glycoscience. Finally, I do hope this *Macro-Glycoligands: Methods and Protocols* book will contribute to the transformation of the discipline of glycoscience from highly specialized research domain to the mainstream biology.

Cleveland, OH, USA ***Xue-Long Sun***

Contents

Preface . . . . *v*
Contributors . . . . *xi*

PART I SYNTHESIS AND CHARACTERIZATION OF GLYCOPOLYMERS

1 Synthesis of Chain-End Functionalized Glycopolymers via Cyanoxyl-Mediated Free Radical Polymerization (CMFRP) 3
Valentinas Gruzdys, Jinshan Tang, Elliot Chaikof, and Xue-Long Sun

2 Protecting-Group-Free Synthesis of Well-Defined Glycopolymers Featuring Negatively Charged Oligosaccharides 13
Luca Albertin

3 Glycopolymers Prepared by Ring-Opening Metathesis Polymerization Followed by Glycoconjugation Using a Triazole-Forming "Click" Reaction. 29
Ronald Okoth and Amit Basu

4 Protecting-Group-Free Synthesis of Glycopolymers and Their Binding Assay with Lectin and Influenza Virus 39
Tomonari Tanaka, Tadanobu Takahashi, and Takashi Suzuki

5 Carbohydrate-Based Initiators for the Cationic Ring-Opening Polymerization of 2-Ethyl-2-Oxazoline 49
Christine Weber, Michael Gottschaldt, Richard Hoogenboom, and Ulrich S. Schubert

6 Heterofunctional Glycopolypeptides by Combination of Thiol-Ene Chemistry and NCA Polymerization. 61
Kai-Steffen Krannig and Helmut Schlaad

7 Preparation of Proteoglycan Mimetic Graft Copolymers 69
Matt J. Kipper and Laura W. Place

PART II GLYCOPOLYMER NANOPARTICLE CONJUGATES

8 Galactosylated Polymer Nano-objects by Polymerization-Induced Self-Assembly, Potential Drug Nanocarriers. 89
Mona Semsarilar, Irene Canton, and Vincent Ladmiral

9 Synthetic Approach to Biotinylated Glyco-Functionalized Quantum Dots: A New Fluorescent Probes for Biomedical Applications 109
Christian K. Adokoh, James Darkwa, and Ravin Narain

10 Surface Modification of Polydivinylbenzene Microspheres with a Fluorinated Glycopolymer Using Thiol-Halogen Click Chemistry. 123
Wentao Song and Anthony M. Granville

11 Glycopolymer-Grafted Polymer Particles for Lectin Recognition 137
Michinari Kohri, Tatsuo Taniguchi, and Keiki Kishikawa

12 Synthesis of Non-spherical Glycopolymer-Decorated Nanoparticles: Combing Thiol-ene with Catecholic Chemistry 149
Xiao Li, Weidong Zhang, and Gaojian Chen

13 Synthetic Approach to Glycopolymer Base Nanoparticle Gold(I) Conjugate: A New Generation of Therapeutic Agents 157
Christian K. Adokoh, James Darkwa, and Ravin Narain

14 Multivalent Glycopolymer-Coated Gold Nanoparticles 169
Sarah-Jane Richards, Caroline I. Biggs, and Matthew I. Gibson

PART III SURFACE IMMOBILIZED GLYCOPOLYMERS

15 Modulation of Multivalent Protein Binding on Surfaces by Glycopolymer Brush Chemistry 183
Kai Yu, A. Louise Creagh, Charles A. Haynes, and Jayachandran N. Kizhakkedathu

16 Oriented Immobilized Sialyloligo-macroligand Microarray 195
Satya Nandana Narla and Xue-Long Sun

17 Glycocalyx Remodeling with Glycopolymer-Based Proteoglycan Mimetics 207
Mia L. Huang, Raymond A.A. Smith, Greg W. Trieger, and Kamil Godula

Index *225*

Contributors

CHRISTIAN K. ADOKOH • *Department of Chemical and Materials Engineering, University of Alberta, Edmonton, AB, Canada; Department of Chemistry, University of Johannesburg, Auckland Park, South Africa*

LUCA ALBERTIN • *Laboratoire de Chimie et Biologie des Métaux, UMR 5249—Université Grenoble Alpes, CEA, CNRS, Grenoble, France*

AMIT BASU • *Department of Chemistry, Brown University, Providence, RI, USA*

CAROLINE I. BIGGS • *Department of Chemistry, The University of Warwick, Coventry, UK*

IRENE CANTON • *The Centre for Stem Cell Biology (CSCB), The University of Sheffield, Sheffield, UK; Department of Biomedical Science, The Centre for Membrane Interactions and Dynamics (CMIAD), The University of Sheffield, Sheffield, UK*

ELLIOT CHAIKOF • *Department of Surgery, Beth Israel Deaconess Medical Center, Harvard Medical School, The Wyss Institute of Biologically Inspired Engineering at Harvard University, Boston, MA, USA*

GAOJIAN CHEN • *Center for Soft Condensed Matter Physics and Interdisciplinary Research, Soochow University, Suzhou, China*

A. LOUISE CREAGH • *Department of Chemical and Biological Engineering, Michael Smith Laboratories,, University of British Columbia, Vancouver, BC, Canada*

JAMES DARKWA • *Department of Chemistry, University of Johannesburg, Auckland Park, South Africa*

MATTHEW I. GIBSON • *Department of Chemistry, The University of Warwick, Coventry, UK*

KAMIL GODULA • *Department of Chemistry and Biochemistry, University of California-San Diego, La Jolla, CA, USA*

MICHAEL GOTTSCHALDT • *Laboratory of Organic and Macromolecular Chemistry (IOMC), Friedrich Schiller University Jena, Jena, Germany; Jena Center for Soft Matter (JCSM), Friedrich Schiller University Jena, Jena, Germany*

ANTHONY M. GRANVILLE • *Centre for Advanced Macromolecular Design, School of Chemical Engineering, The University of New South Wales, Sydney, NSW, Australia*

VALENTINAS GRUZDYS • *Department of Chemistry, Chemical and Biomedical Engineering and Center for Gene Regulation in Health and Disease (GRHD), Cleveland State University, Cleveland, OH, USA*

CHARLES A. HAYNES • *Department of Chemical and Biological Engineering, Michael Smith Laboratories, Michael Smith Laboratories, University of British Columbia, Vancouver, BC, Canada*

RICHARD HOOGENBOOM • *Supramolecular Chemistry Group, Department of Organic and Macromolecular Chemistry, Ghent University, Ghent, Belgium*

MIA L. HUANG • *Department of California-San Deigo, of California-San Diego, La Jolla, CA, USA*

MATT J. KIPPER • *Department of Chemical and Biological Engineering and School of Biomedical Engineering, Colorado State University, Fort Collins, CO, USA*

KEIKI KISHIKAWA • *Division of Applied Chemistry and Biotechnology, Graduate School of Engineering, Chiba University, Chiba, Japan*

JAYACHANDRAN N. KIZHAKKEDATHU • *Department of Pathology and Laboratory Medicine, Centre for Blood Research, University of British Columbia, Vancouver, BC, Canada; Department of Chemistry, University of British Columbia, Vancouver, BC, Canada*
MICHINARI KOHRI • *Division of Applied Chemistry and Biotechnology, Graduate School of Engineering, Chiba University, Chiba, Japan*
KAI-STEFFEN KRANNING • *Department of Colloid Chemistry, Max Planck Institute of Colloids and Interfaces, Potsdam, Germany*
VINCENT LADMIRAL • *ICGM (Institut Charles Gerhardt) UMR 5253 (CNRS-ENSCM-UM), Université de Montpellier, Montpellier, France*
XIAO LI • *Center for Soft Condensed Matter Physics and Interdisciplinary Research, Soochow University, Suzhou, China*
RAVIN NARAIN • *Department of Chemical and Materials Engineering, University of Alberta, Edmonton, AB, Canada*
SATYA NANDANA NARLA • *Department of Chemistry, Chemical and Biomedical Engineering, Center for Gene Regulation of Health and Disease (GRHD), Cleveland State University, Cleveland, OH, USA*
RONALD OKOTH • *Department of Chemistry, Brown University, Providence, RI, USA*
LAURA W. PLACE • *Department of Chemical and Biological Engineering and School of Biomedical Engineering, Colorado State University, Fort Collins, CO, USA*
SARAH-JANE RICHARDS • *Department of Chemistry, The University of Warwick, Coventry, UK*
HELMUT SCHLAAD • *Institute of Chemistry, University of Potsdam, Potsdam, Germany*
ULRICH S. SCHUBERT • *Laboratory of Organic and Macromolecular Chemistry (IOMC), Friedrich Schiller University Jena, Jena, Germany; Jena Center for Soft Matter (JCSM), Friedrich Schiller University Jena, Jena, Germany*
MONA SEMSARILAR • *IEM (Institut Européen des Membranes), UMR 5635 (CNRS-ENSCM-UM), Université de Montpellier, Montpellier, France*
RAYMOND A.A. SMITH • *Department of Chemistry and Biochemistry, University of California-San Deigo, La Jolla, CA, USA*
WENTAO SONG • *Centre for Advanced Macromolecular Design, School of Chemical Engineering, The University of New South Wales, Sydney, NSW, Australia*
XUE-LONG SUN • *Department of Chemistry, Chemical and Biomedical Engineering and Center for Gene Regulation in Health and Disease (GRHD), Cleveland State University, Cleveland, OH, USA*
TAKASHI SUZUKI • *Department of Biochemistry, School of Pharmaceutical Sciences, University of Shizuoka, Suruga-Ku, Shizuoka, Japan*
TADANOBU TAKAHASHI • *Department of Biochemistry, School of Pharmaceutical Sciences, University of Shizuoka, Suruga-Ku, Shizuoka, Japan*
TOMONARI TANAKA • *Department of Biobased Materials Science, Graduate School of Science and Technology, Kyoto Institute of Technology, Matsugasaki, Sakyo-Ku, Kyoto, Japan*
JINSHAN TANG • *Department of Chemistry, Chemical and Biomedical Engineering and Center for Gene Regulation in Health and Disease (GRHD), Cleveland State University, Cleveland, OH, USA*
TATSUO TANIGUCHI • *Division of Applied Chemistry and Biotechnology, Graduate School of Engineering, Chiba University, Chiba, Japan*
GREG W. TRIEGER • *Department of Chemistry and Biochemistry, University of California-San Diego, La Jolla, CA, USA*

CHRISTINE WEBER • *Laboratory of Organic and Macromolecular Chemistry (IOMC), Friedrich Schiller University Jena, Jena, Germany; Jena Center for Soft Matter (JCSM), Friedrich Schiller University Jena, Jena, Germany*
KAI YU • *Department of Pathology and Laboratory Medicine, Centre for Blood Research, University of British Columbia, Vancouver, BC, Canada*
WEIDONG ZHANG • *Center for Soft Condensed Matter Physics and Interdisciplinary Research, Soochow University, Suzhou, China*

Part I

Synthesis and Characterization of Glycopolymers

Chapter 1

Synthesis of Chain-End Functionalized Glycopolymers via Cyanoxyl-Mediated Free Radical Polymerization (CMFRP)

Valentinas Gruzdys, Jinshan Tang, Elliot Chaikof, and Xue-Long Sun

Abstract

Glycopolymers are often used as glyco-macroligands for biological research and biomedical applications in carbohydrate recognitions. Chain-end functionalized glycopolymers show more potential for practical applications, such as protein modification and solid-phase bioassays. In particular, the chain-end group allows for direct one-to-one attachment or facilitates site-specific and oriented immobilization onto solid surfaces. A series of derivatized arylamine initiators are used to generate chain-end functionalized glycopolymers by cyanoxyl-mediated free radical polymerization (CMFRP). Important features of this strategy include the capacity to produce polymers of low polydispersity (PDI <1.5) under aqueous conditions using unprotected monomers bearing a wide range of functional groups. In addition, it provides a one-pot method to synthesize α,ω-telechelic glycopolymers with derivatized arylamine at one site and *O*-cyanate at the other site. In the process, the capacity to orthogonally label glycopolymers or otherwise conjugate them to proteins and other molecules is greatly enhanced.

Key words Glycopolymer, Chain-end functionalized, Carbohydrate, Cyanoxyl-mediated free radical polymerization, Lactose, Biotin

1 Introduction

Glycopolymers, namely polymers with multivalent carbohydrate pendant groups, have been extensively explored for different biological research and applications for decades [1–8]. It is generally accepted that synthetic glycopolymers can mimic the functions of naturally occurring carbohydrates [9, 10] and have been employed as cell-surface receptors for studying functions of biologically active carbohydrates and carbohydrate-binding proteins [11]. Recently, the potential utility of glycopolymers in bio- and immunochemical assays as biocapture reagents and for microarray applications has been demonstrated by the addition of functional anchor groups either as pendants to the polymer backbone or at the chain end [12]. The chain-end group of the glycopolymer is of particular focus for bio-functionalization, as it allows for a direct one-to-one

Xue-Long Sun (ed.), *Macro-Glycoligands: Methods and Protocols*, Methods in Molecular Biology, vol. 1367,
DOI 10.1007/978-1-4939-3130-9_1,

attachment [13] or facilitates site-specific and oriented immobilization onto solid surfaces [14]. The latter is especially exploited for chip-based bioassays and cell-adhesion studies [15, 16] as they mimic the 3D display of carbohydrates on the cell surfaces.

Generating chain-end functionalized glycopolymers with oligosaccharide units of increasing functional complexity poses a number of significant challenges, including a requirement for serial protection/deprotection steps or further polymer derivatization after initial synthesis. We have previously reported that cyanoxyl-mediated free radical polymerization (CMFRP) of unprotected glycomonomers can be conducted in an aqueous solution, is tolerant to a broad range of functional groups including -OH, $-NH_2$, -COOH, and SO_3 moieties, and can yield glycopolymers with low polydispersity (PDI <1.5) [17–19]. Here, we describe a straightforward approach to synthesize chain-end functionalized glycopolymers using functionalized arylamine initiators in conjunction with CMFRP (Fig. 1) [20]. In addition, it provides a one-pot method to synthesize α,ω-telechelic glycopolymers with derivatized arylamine at one site and *O*-cyanate at the other site. The *O*-cyanate chain-end group can be used for site-specific conjugation [21] or immobilization [22–24] onto amine-containing molecules via isourea bond formation in mild conditions.

2 Materials

2.1 Chemicals

1. *P*-Anisidine(4-methoxyaniline).
2. Acrylamide (AM).
3. 2-(4-Aminophenyl) ethylamine.

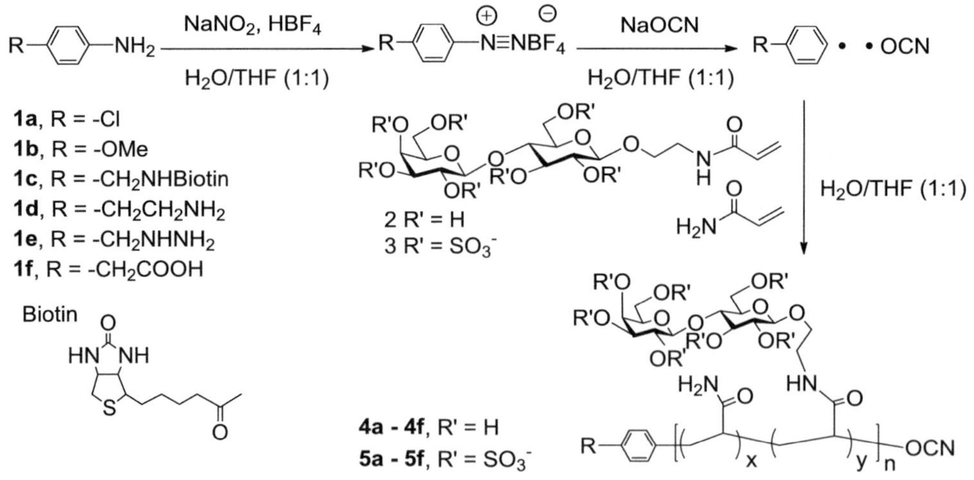

Fig. 1 Synthesis of chain-end functionalized glycopolymers via cyanoxyl-mediated free radical polymerization. Reprinted with permission from *Bioconjugate Chem.*, Vol. 15, No. 5, 2004. Copyright 2015 American Chemical Society

4. 4-Aminobenzoic hydrazide.
5. 4-Aminophenylacetic acid.
6. Celite.
7. 4-Chloroaniline.
8. Tetrafluoroboric acid (48 % aqueous solution).
9. N-(9-fluorenymethoxycarbonyloxy) succinimide.
10. 4′-Hydroxyazo-benzene-2-carboxylic acid (HABA).
11. *N*-hydroxysuccinimide-biotin (Biotin-NHS).
12. *p*-Nitrobenzylamine.
13. Pyridine.
14. Palladium-carbon (Pd/C).
15. Sodium nitrite ($NaNO_2$).
16. Sodium cyanate (NaOCN).
17. Triethylamine (Et_3N).
18. Glycomonomer lactosyl acrylamide was synthesized as previously described [17].
19. Glycomonomer-sulfated lactosyl acrylamide was synthesized as previously described [17].

2.2 Solvents

1. Deionized water (DI H_2O).
2. Deuterated water (D_2O).
3. Dimethylformamide (DMF).
4. Methanol (MeOH).
5. Tetrahydrofuran (THF).
6. 0.1 M PBS buffer, pH 7.4.

3 Methods

CMFRP is a straightforward approach to synthesize chain-end functionalized glycopolymers with functionalized arylamine as an initiator, which excludes protection/deprotection and conjugation steps often used in other polymerization methods. For the arylamine initiator, commercially available 4-chloroaniline, 2-(4-aminophenyl) ethylamine, 4-aminobenzoic hydrazide, and 4-aminophenylacetic acid were used for amine, hydrazide, and carboxylate chain-end glycopolymer synthesis. 4-Aminobenzylbiotinamide was synthesized for biotin chain-end glycopolymer synthesis as shown in Fig. 2. As model glycomonomers, lactosyl and sulfated lactosyl acrylamide were synthesized as previously described [17]. Finally, the *O*-cyanate chain-end of the polymer could be converted to hydroxyl group by treating it with pyridine in water quantitatively (Fig. 3).

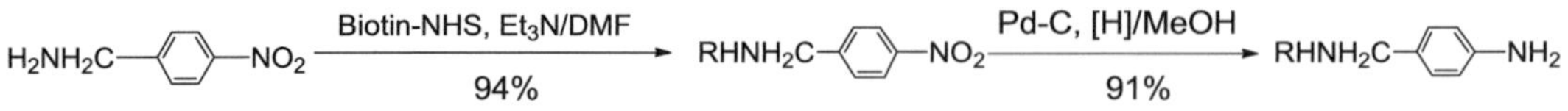

Fig. 2 Synthesis of biotin arylamine derivative

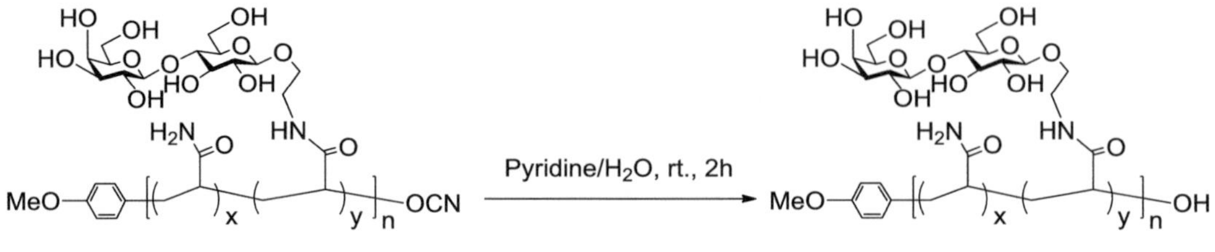

Fig. 3 Conversion of terminal cyanate (OCN) to hydroxyl group. Reprinted with permission from *Bioconjugate Chem.*, Vol. 15, No. 5, 2004. Copyright 2015 American Chemical Society

3.1 Synthesis of Biotin Arylamine Derivatives

3.1.1 Synthesis of 4-Nitrobenzyl-Biotinamide (2)

1. Add triethylamine (Et_3N) (1.0 mL) into a solution of *p*-nitrobenzylamine (144 mg, 0.78 mmol) in DMF (5 mL).
2. Stir the solution for 30 min at room temperature.
3. Add N-hydroxysuccinimide-biotin (200 mg, 0.56 mmol) into the solution above.
4. Stir the reaction mixture for 24 h at room temperature.
5. Concentrate the reaction mixture under vacuum to give a residue.
6. Purify the residue by silica gel column using chloroform and methanol (8:1, v/v) as eluent to afford **2** (184 mg, 94 %).
7. Characterize compound **2** by ^{1}H NMR and MS spectrometry: ^{1}H NMR signals are observed ($CDCl_3/CD_3OD$) δ: 8.08 (d, 2 H, *J*=9.9 Hz), 7.53 (d, 2 H, *J*=9.9 Hz), 4.44 (m, 1 H), 4.26 (dd, 1 H, *J*=4.5, 8.3 Hz), 3.69 (t, 1 H, *J*=4.5 Hz), 3.33 (t, 1 H, *J*=5.8 Hz), 3.14 (m, 1 H), 2.90 (dd, 1 H, *J*=12.4, 4.5 Hz). 2.67 (s, 2 H), 2.26 (t, 2 H, *J*=7.9 Hz), 1.73 (m, 4 H), 1.46–1.38 (m, 3 H). HR-MS (EI) is calculated for $C_{17}H_{22}N_4O_4SLi$ 385.1522 and found 385.1533 $[M+Li]^+$.

3.1.2 Synthesis of 4-Aminobenzyl-Biotinamide (3)

1. In the presence of palladium on carbon (Pd-C) (40 mg), charge compound **2** (100 mg, 0.264 mmol) in methanol (5 mL) with hydrogen balloon for hydrogenation for 4 h at room temperature (*see* **Note 1**).
2. Filter the reaction mixture, collect the filtrate and concentrate it under evaporator to provide a residue.
3. Purify the residue by silica gel column using chloroform and methanol (5:1, v/v) as eluent to afford **3** (82 mg, 91 %) (*see* **Note 1**).
4. Characterize compound **3** by ^{1}H NMR and MS spectrometry: ^{1}H NMR signals are observed (CDCl3/CD3OD) δ: 7.13

(d, 2 H, J=9.8 Hz), 6.78 (d, 2 H, J=9.8 Hz), 4.59 (m, 1 H), 4.39 (m, 1 H), 4.33 (1 H, s), 3.24 (m, 1 H), 3.00 (dd, 1 H, J=4.9, 12.7 Hz), 2.80 (d, 1 H, J=12.7 Hz), 2.30 (t, 2 H, J=9.8 Hz), 1.84–1.60 (m, 4 H), 1.58–1.46 (m, 3 H). HR-MS (EI) is calculated for C17H24N4O2SLi 355.1780 and found 355.1780 [M+Li]+.

3.2 Cyanoxyl-Mediated Free Radical Polymerization of Acrylamide-Derived Glycomonomers with Acrylamide Initiated by $RC_6H_4N{\equiv}N^+BF_4^-$/ NaOCN

1. In a three-neck flask, add arylamine (6.03×10^{-2} mmol) and HBF_4 (17 mg, 9.04×10^{-2} mmol, 48 wt % aqueous solution) followed by deionized water (DI H_2O)/THF (1 mL, 1:1 (v/v)) and dissolve them well.
2. Seal the flask and replace the air with argon (Ar) and keep it at 0 °C under argon (Ar) atmosphere (Fig. 1) (*see* **Note 2**).
3. Afterwards, add sodium nitrite ($NaNO_2$) (5 mg, 7.2×10^{-2} mmol) in DI water (0.5 mL) to the reaction medium to generate the diazonium salt $RC_6H_4N{\equiv}N^+BF_4^-$ for 30 min at 0 °C under argon (Ar) atmosphere (*see* **Notes 2** and **3**).
4. Transfer a degassed solution of glycomonomer (**2**/**3**) (6.03×10^{-1} mmol), acrylamide (2.41×10^{-3} mol), and sodium cyanate (NaOCN) (4 mg, 6.03×10^{-2} mmol) dissolved in 0.5 mL of DI H_2O into the flask containing the diazonium salt (*see* **Notes 4** and **5**).
5. Heat the polymerization solution to 65 °C in oil bath and stir it for 16 h.
6. Quench the polymerization by opening flask and exposing reaction to air.
7. Evaporate the reaction solution under vacuum to remove THF solvent and then transfer the aqueous solution into a dialysis tube (3500 Da M_W cutoff) for dialysis for 2 days at room temperature to remove inorganic salt and impurities.
8. Lyophilize the dialysis solution to yield the glycocopolymer (**4**/**5**).
9. Calculate the conversion yield by weight for the resultant glycopolymer.
10. Determine the carbohydrate content and other components from ^{1}H NMR spectrum (Fig. 4).

3.3 Conversion of Cyanate (OCN) of Glycopolymer 4b to a Hydroxyl End Group

1. Add pyridine (0.5 mL) into a solution of glycopolymer **4b** (16 mg, 2.1×10^{-3} mmol) in DI H_2O (2 mL).
2. Stir the mixture at room temperature for 2 h, followed by dialysis against water at room temperature for 2 days to remove excess pyridine and glutaconaldehyde.
3. Lyophilize the dialysis solution to yield the glycopolymer **6b** (16 mg, quantitatively).

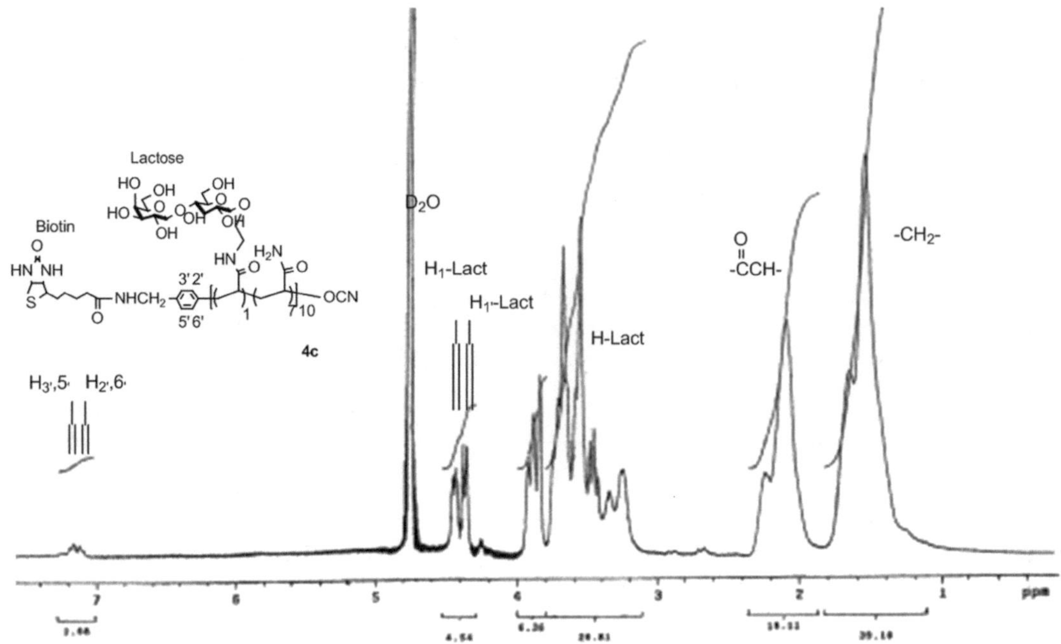

Fig. 4 ^{1}H NMR spectrum of biotin chain-end functionalized glycopolymer **4c** in D_2O. Reprinted with permission from *J. AM. CHEM. SOC.* 2002, 124, 7258–7259. Copyright 2015 American Chemical Society

3.4 ^{1}H NMR Characterization of Glycopolymers

The presence of a terminal phenyl group in the resultant polymer allows for easy determination of carbohydrate density and average molecular weight of the glycopolymer by comparing the integration of phenyl protons with that of sugar anomeric and polymer backbone protons in ^{1}H NMR spectrum. As shown in Fig. 4, comparison of the integrated signals produced from the chain-end phenyl protons ($H_{2',6'}$ and $H_{3',5'}$) with those due to the anomeric protons of lactose ($H_{1'\text{-Lac}}$ and $H_{1\text{-Lac}}$) and the backbone protons (-COCH-, $-CH_2-$) indicate that the phenyl chain-end functionalized glycopolymer **4c** has 10 lactose units and 70 acrylamide units on average.

3.5 Complex Formation of Biotin Chain-End Functionalized Glycopolymer with Streptavidin

High-affinity binding of biotin to streptavidin (*affinity constant* 10^{13-15} M^{-1}) has led to the use of streptavidin as a molecular adapter in diverse applications [25]. The ability of biotin-glycopolymers (**4**/**5**) to specifically bind streptavidin was assessed using a HABA-streptavidin assay [25]. HABA (λ_{max} 350 nm) changes color from yellow to red (λ_{max} 500 nm) upon binding to streptavidin (Fig. 5, trial C). When HABA is added to a solution of streptavidin saturated with free biotin or biotin glycopolymers **4c** and **5c**, a red shift is not observed (Fig. 5, trials D, E, and F). In contrast, a color change is noted when HABA is added to a solution of streptavidin

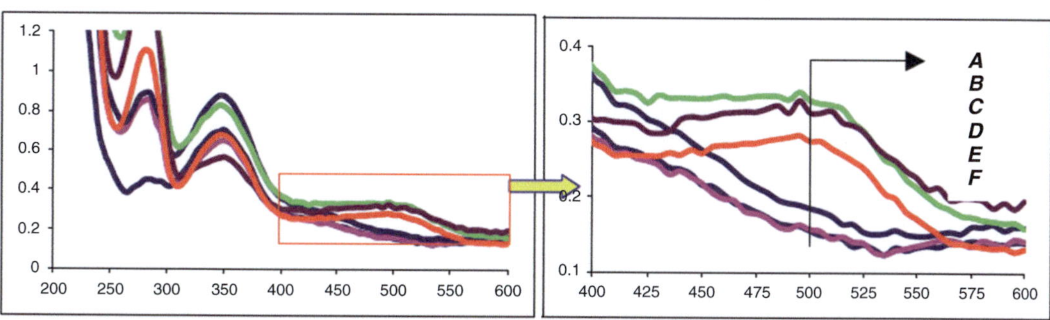

Fig. 5 Monitoring of streptavidin-biotin-glycopolymer (**4c/5c**) interactions by UV-vis spectroscopy: (*A*) streptavidin + **4a** + HABA; (*B*) streptavidin + **5a** + HABA; (*C*) streptavidin + HABA; (*D*) streptavidin + biotin + HABA; (*E*) streptavidin + **4c** + HABA; (*F*) streptavidin + **5c** + HABA. Reprinted with permission from *Bioconjugate Chem.*, Vol. 15, No. 5, 2004. Copyright 2015 American Chemical Society

with non-biotin-containing glycopolymers **4a** and **5a** (Fig. 2, trials A and B). These results demonstrate glycopolymer-protein hybrid formation through specific binding of chain-end biotin of glycopolymers **4c** and **5c** to streptavidin. In principle, streptavidin has four free biotin-binding sites. However, the HABA assay reveals an average occupancy of 3.0–3.6 glycopolymer chains per streptavidin molecule. Full occupancy may be limited as a result of steric factors.

1. Dissolve biotin chain-end functionalized glycopolymer **4c/5c** (1.33×10^{-4} mmol) and streptavidin (1.67×10^{-5} mmol) in 0.1 M PBS (pH 7.4, 0.2 mL) and incubate at room temperature for 2 h.
2. Take UV–Vis spectroscopy before and after adding 4′-hydroxyazo-benzene-2-carboxylic acid (HABA, 1.33×10^{-4} mmol, Fig. 5).

4 Notes

1. Pd/C frequently ignites when it first comes in contact with methanol and represents a significant safety risk. The following procedures are recommended whenever Pd/C is used in conjunction with hydrogen gas balloon.

   Set up reaction:

   (a) Vacuum out the reaction vessel (it should have at least two openings that can be closed or opened selectively) and backfill with an inert gas (nitrogen or argon).

   (b) Weigh out the desired amount of Pd/C and transfer into the reaction flask under an inert atmosphere.

(c) Carefully add methanol by creating a stream down the side of the flask wall.

(d) Add the reaction substrate in methanol solution.

(e) Begin stirring the reaction mixture, then vacuum out the flask just until the solvent begins to bubble, and then carefully backfill with inert gas.

(f) Repeat **step (e)** twice more.

(g) Attach a balloon of hydrogen to your flask with an adapter that allows the balloon to be closed off from the reaction flask.

(h) With the hydrogen balloon closed off, vacuum out the flask until the solvent begins to bubble, and then open the balloon to the flask.

(i) Repeat **step (h)** twice more.

Work-up:

(a) Detach the hydrogen balloon from the flask and fill it with inert gas.

(b) Filter the reaction mixture through a bed of Celite.

(c) Taking care not to let the filter cake to dryness, wash with the desired solvent (typically the same solvent used in the reaction).

(d) Disconnect the filter from the receiving flask, and then add several mL of water to the filter.

(e) Discard the slurried Pd/C and filter aid in a dedicated waste jar that contains water.

2. In CMFRP, all steps need to be taken under Ar atmosphere. Reaction container is degassed under vacuum and backfilled with Ar before addition of solvent.

3. In CMFRP, diazonium formation from arylamine initiator can be verified by the yellowish appearance of the solution, often before the addition of HBF_4, with subsequent darkening observed.

4. Glycomonomer, acrylamide, and NaOCN solution in DI water must be degassed before introducing into the polymerization chamber by freeze-pump-thaw degassing procedure. Briefly, the solution in a sealed Schlenk flask is frozen by immersion of the flask in liquid N_2. When the solvent is completely frozen, the flask is opened to high vacuum and pumped for 2–3 min, with the flask still immersed in liquid N_2. The flask is then closed and warmed until the solvent has completely melted. This process is repeated at least three times and after the last cycle the flask is backfilled with Ar gas.

5. The degassed solution containing glycomonomer, acrylamide, and NaOCN is transferred into the polymerization chamber by connecting the two with a double ended needle followed by the negative pressure (as a result of vacuum) inside the polymerization chamber, which is then backfilled with Ar gas again.

Acknowledgments

This work was supported by grants from the NIH and NSF. The authors acknowledge the Emory University NMR and Mass Spectrometry Centers for use of their facilities.

References

1. Lundquist JJ, Toone EJ (2002) The cluster glycoside effect. Chem Rev 102:555–578
2. Armes RN, Steven P (2003) Synthesis and aqueous solution properties of novel sugar methacylate-based homopolymers and block polymers. Biomacromolecules 4:1746–1758
3. Wulff G, Schmid J, Venhoff T (1996) The synthesis of polymerizable vinyl sugars. Macromol Chem Phys 197:259–274
4. Okada M (2001) Molecular design and syntheses of glycopolymers. Prog Polym Sci 26:67–104
5. Narian R, Jhurry D, Wulff G (2002) Synthesis and characterization of polymers containing linear sugar moieties as side groups. Eur Polym J 38:273–280
6. Ohno K, Izu Y, Yamamoto S, Miyamoto T, Fukuda T (1999) Nitroxide-controlled free radical polymerization of a sugar-carrying acryloyl monomer. Macromol Chem Phys 200:1619–1625
7. Strong LE, Kiessling LL (1999) A General synthetic route to defined, biologically active multivalent arrays. J Am Chem Soc 121:6193–6196
8. Roy R, Laferrière CA, Pon RA, Gamian A (1994) Induction of rabbit immunoglobulin G antibodies against synthetic sialylated neoglycoproteins. Methods Enzymol 247:351–361
9. Ting SSR, Chen G, Stenzel MH (2010) Synthesis of glycopolymers and their multivlent recognition with lectins. Polym Chem 1:1392–1412
10. Voit B, Appelhans D (2010) Glycopolymers of various architectures: more than mimicking nature. Macromol Chem Phys 211:727–735
11. Boyer C, Bulmus V, Liu J, Davis TP, Stenzel MH, Barner-Kowollik C (2007) Well-defined protein-polymer conjugates via in situ RAFT polymerization. J Am Chem Soc 129: 7145–7154
12. Narla SN, Nie H, Li Y, Sun X-L (2012) Recent advances in synthesis and biomedical applications of chain-end functionalized glycopolymers. J Carbohydrate Chem 31:67–92
13. Roth PJ, Jochum FD, Zentel R, Theato P (2010) Synthesis of hetero-telechelic α, ω biobunctionalized polymers. Biomacromolecules 11:238–244
14. Gestwicki JE, Cairo CW, Mann DA, Owen RM, Kiessling LL (2002) Selective immobilization of multivalent ligands for surface plasmon resonance and fluorescence microscope. Anal Biochem 305:149–155
15. Tugulu S, Silacci P, Stergiopulos N, Klok H-A (2007) RGD-Functionalized polymer brushes as substrates for the integrin specific adhesion of human umbilical vein endothelial cells. Biomaterials 28:2536–2546
16. Hasegawa T, Kondoh S, Matsuura K, Kobayashi K (1999) Rigid helical poly(glycosyl phenyl isocyanide)s: synthesis, conformational analysis, and recognition by lectins. Macromolecules 32:6595–6603
17. Sun X-L, Grande D, Baskaran S, Hanson SR, Chaikof EL (2002) Glycosaminoglycan mimetic biomaterials. 4. Synthesis of sulfated lactose-based glycopolymers that exhibit anticoagulant activity. Biomacromolecules 3:1065–1070
18. Sun X-L, Faucher KM, Houston M, Grande D, Chaikof EL (2002) Design and synthesis of biotin chain-terminated glycopolymers for surface glycoengineering. J Am Chem Soc 124:7258–7259
19. Faucher KM, Sun X-L, Chaikof EL (2003) Fabrication and characterization of glycocalyx-mimetic surfaces. Langmuir 19:1664–1670

20. Hou S, Sun X-L, Dong CM, Chaikof EL (2004) Facile synthesis of chain-end functionalized glycopolymers for site-specific bioconjugation. Bioconjugate Chem 15:954–959
21. Gruzdys V, Zhang H, Sun X-L (2014) Glycomodification of protein with *O*-cyanate chain-end functionalized glycopolymer via isourea bond formation. J Carbohy Chem 33:1–13
22. Narla SN, Sun X-L (2011) Orientated glycomacroligand formation based on site-specific immobilization of *O*-cyanate chain-end functionalized glycopolymer. Org Biomol Chem 9:845–850
23. Narla SN, Sun X-L (2012) Immobilized sialyloligo-macroligand and its protein binding specificity. Biomacromolecules 13:1675–1682
24. Narla SN, Sun X-L (2012) Glyco-macroligand microarray with controlled orientation and glycan density. Lab Chip 12:1656–1663
25. Weber PC, Wendoloski JJ, Pantoliano MW, Salemme FR (1992) Crystallographic and thermodynamic comparison of natural and synthetic ligands bound to streptavidin. J Am Chem Soc 114:3197

Chapter 2

Protecting-Group-Free Synthesis of Well-Defined Glycopolymers Featuring Negatively Charged Oligosaccharides

Luca Albertin

Abstract

Control of the macromolecular architecture is essential to enable sophisticated functions for glycopolymers and to allow a precise correlation between these functions and the polymer structure. A number of biologically important ligands are negatively charged oligosaccharides that are difficult to manipulate in organic solvent and that are hardly amenable to protection/deprotection strategies. RAFT polymerization is a simple and robust technique that enables the synthesis of well-defined glycopolymers directly in aqueous solution and starting from unprotected vinyl glycomonomers. Here I describe how RAFT polymerization can be combined with reductive amination to transform negatively charged oligosaccharides having 5–20 monosaccharide units into well-defined glycopolymers directly in water and without the need to resort to protecting-group chemistry.

Key words Glycopolymers, RAFT, Radical polymerization, Glycuronan, (1→4)-α-L-guluronan, (1→4)-β-mannuronan

1 Introduction

Carbohydrates of varying complexity are present in all cells and in numerous biological macromolecules, where they usually decorate the outer surface. Hence they are ideally situated to mediate or modulate a variety of cell-cell, cell-matrix, and cell-molecule interactions which are critical to the development and function of a complex multicellular organism. Moreover, they can mediate the interaction between different organisms, such as that between a host and a parasite or symbiont. Hence the interest for glycopolymer architectures, i.e., synthetic polymers possessing a non-carbohydrate main chain but featuring pendant and/or terminal carbohydrate moieties, as a tool in glycobiology research.

Besides the presence of the appropriate carbohydrate(s), control of the macromolecular architecture is essential to enable

Xue-Long Sun (ed.), *Macro-Glycoligands: Methods and Protocols*, Methods in Molecular Biology, vol. 1367, DOI 10.1007/978-1-4939-3130-9_2, © Springer Science+Business Media New York 2016

sophisticated functions for glycopolymers [1–3] and to allow a precise correlation between these functions and the polymer structure. For this reason, over the past 20 years a trend has emerged in which more and more polymer chemists got involved in the synthesis of novel glycopolymers via precise polymerization techniques, while a greater number of biochemists and carbohydrate chemists adopted these techniques for designing tailored glycoligands. As a result, a rich literature is now available on the synthesis of well-defined glycopolymers carrying mono-, di-, or trisaccharides [4, 5]. Only a few reports deal instead with glycoconjugates featuring carbohydrates of higher complexity [6, 7], and (or) with negatively charged carbohydrates [8–13].

Negatively charged oligosaccharides include a number of biologically important ligands such as sialyl oligosaccharides and low molar mass glycosylaminoglycans [14]. Due to their size and the presence of several functional groups (e.g., carboxylic and sulphate), these carbohydrates are difficult to manipulate in organic solvent [15] and their protecting-group chemistry is exceedingly time consuming [16]. In this context, we have suggested a water-based approach to the synthesis of well-defined glycopolymers featuring negatively charged oligosaccharides with 5–20 monosaccharide units [17, 18]. The glycomonomers are prepared by reductive amination of the free oligosaccharides (eventually) followed by acylation of the resulting 1-amino-1-deoxyalditol directly in water and without the need to resort to protective groups' chemistry, as described in Fig. 1 (*see* also **Note 1**).

They are then copolymerized with *N*-(2-hydroxyethyl)methacrylamide (HEMAm; *see* **Note 2**) by aqueous reversible addition-fragmentation chain transfer polymerization (RAFT, *see* **Note 3**) [19] to afford well-defined poly(HEMAm-*graft*-oligosaccharide) glycopolymers carrying a thiocarbonylthio moiety at their ω-end (Fig. 2). The latter can be easily converted into a thiol [20] to enable further conjugation or grafting to a gold surface [21]. It is worth noting that although these protocols were developed for (1→4)-α-L-guluronan and (1→4)-β-D-mannuronan with M_n = 1000–4000 Da, they can be safely applied to low molar mass glycosylaminoglycans and any other oligo/polysaccharide with a reducing end (*see* also **Note 4**).

2 Materials

2.1 Chemicals

1. Deionized water is produced in-house with a MilliQ apparatus (Millipore) and used for all experiments.
2. *N*-(2-Hydroxyethyl)methacrylamide (HEMAm) is synthesized from methacryloyl chloride and 2-aminoethyl alcohol as in [18] (*see* **Notes 5** and **6**).

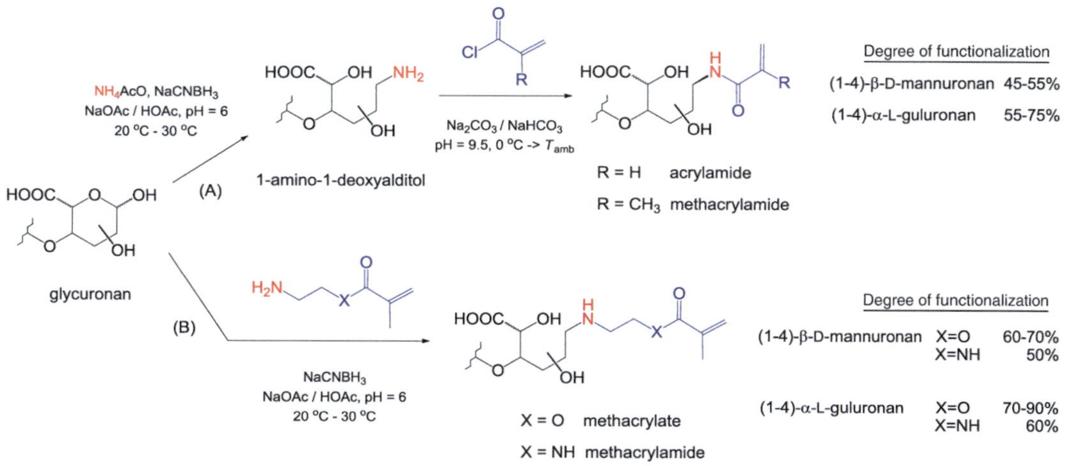

Fig. 1 Synthesis of glycuronan glycomonomers in aqueous solution by reductive amination: (**a**) in two steps via a 1-amino-1-deoxyalditol or (**b**) in one step using an excess of ethylenic monomer carrying a primary amino function. Note that the two strategies give access to glycomonomers with different polymerizable ethylenic moieties

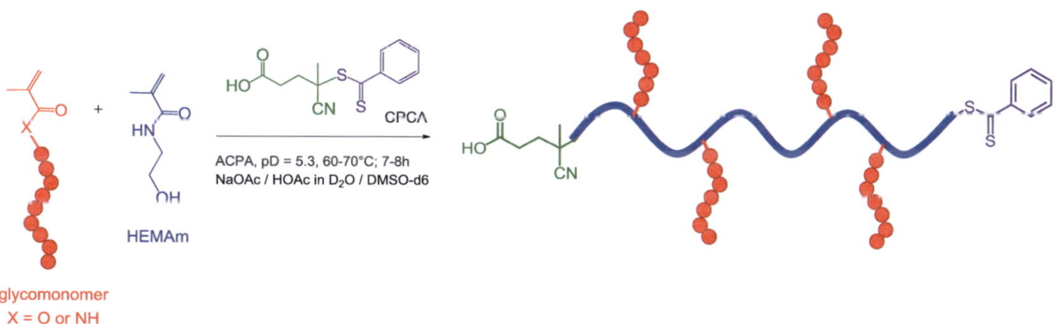

Fig. 2 RAFT copolymerization of oligosaccharide-derived glycomonomers with HEMAm and structure of the resulting glycopolymers

3. 4-Cyano-4-[(phenylcarbonothioyl)sulfanyl] pentanoic acid (CPCA) is prepared from 4,4′-azobis(cyanopentanoic acid) (ACPA) and bis(thiobenzoyl) disulfide according to the published method [22] (*see* **Note 7**).
4. (1 → 4)-β-D-Mannuronan and (1 → 4)-α-L-guluronan oligosaccharides, sodium salt (Elicityl SA, Crolles, France).
5. 0.11 M Acetate buffer (pH 5.9) is prepared by introducing 84.16 g of NaOAc and 423 μL (444 mg) of HOAc in a graduated cylinder and by filling it up to 1.0 L with H_2O.
6. 1.0 M Deuterated acetate buffer (pD 5.2) is prepared by dissolving 601 mg of NaOAc and 153 μL (160 mg) of HOAc in 10 mL of D_2O.

7. 0.60 M Sodium carbonate buffer (pH 9.5) is prepared by dissolving 44.52 g of $NaHCO_3$ and 7.42 g of Na_2CO_3 in ~900 mL of H_2O, by adjusting the pH with 1 M NaOH, and by filling up to 1.0 L.

2.2 Commercial Reagents

1. 2-Aminoethyl methacrylate hydrochloride (AEM·HCl, 90 %, Aldrich; *see* **Notes 8** and **9**).
2. Ammonium acetate (NH_4OAc).
3. 4,4′-Azobis(cyanopentanoic acid) (ACPA).
4. 2,6-Di-tert-butyl-4-methyl phenol (BHT).
5. DMSO-d6.
6. D_2O.
7. Methacryloyl chloride (distilled, ≥97 % Fluka).
8. Sodium cyanoborohydride ($NaBH_3CN$).

2.3 Chromatography

2.3.1 Thin-Layer Chromatography

1. Thin-layer chromatography (TLC) analyses are performed on aluminum-backed silica gel plates (60 Å, 15 μm, Merck).
2. Following solvent evaporation, the developed plates are exposed to a UV lamp ($\lambda = 254$ nm) for spot detection.

2.3.2 Flash Chromatography

1. Flash chromatography is carried out with a glass column (Ø 7 cm).
2. The column is packed with ~24 cm of silica gel from Merck (60 Å, 40–60 μm).

2.3.3 Analytical Size-Exclusion Chromatography

1. Analytical size-exclusion chromatography (SEC) of the glycopolymers is carried out at 30 °C using a bench of two Shodex columns (SB-802 HQ and SB-803 HQ, 300 × 8 mm) and a guard column (50 × 6 mm); mobile phase 0.1 M $NaNO_3$, 0.03 % w/v NaN_3, 0.01 M EDTA; flow rate 0.5 mL/min; injection volume 108 μL. The system is equipped with a differential refractometer and a multi-angle laser light scattering (MALLS) detector.
2. Data are analyzed with ASTRA 5.3 software (Wyatt Technology Corp.) using the Zimm model and linear fitting.
3. Differential refractive index increments (dn/dc) for the copolymers are estimated from the mass fraction (F_m) of each monomer and the dn/dc of the corresponding homopolymer according to the formula [23]

$$\mathrm{d}n/\mathrm{d}c = F_{\mathrm{m},1}(\mathrm{d}n/\mathrm{d}c)_1 + F_{\mathrm{m},2}(\mathrm{d}n/\mathrm{d}c)_2 \quad (1)$$

To this end, values of 0.165 and 0.208 mL/g are used for glycuronans [24] and poly(HEMAm) [18], respectively.

4. Samples are prepared by diluting the polymerization mixture or by dissolving pure polymer samples directly in the mobile phase ($c \cong 4$ g/L) and by filtering through a 0.22 μm sterile syringe filter.

2.4 Nuclear Magnetic Resonance

1. Resonance frequency of 400.13 and 100.62 MHz for ^1H and ^{13}C nuclei, respectively.
2. ^1H NMR: 90° pulses, acquisition time 3 s, pulse sequence recycle delay 9 s.
3. Sodium 3-(trimethylsilyl)propanoate (TSP) or sodium 3-(trimethylsilyl)propane-1-sulfonate (DSS) are used as an internal reference for samples dissolved in D_2O, whereas tetramethylsilane (TMS) is used in all other cases. Chemical shifts (δ/ppm) are referenced to $\delta_{TSP} = -0.017$ ppm (^1H) and −0.149 ppm (^{13}C); or to δ_{DSS} and $\delta_{TMS} = 0.000$ ppm (^1H and ^{13}C).

2.5 Diafiltration

1. Glycomonomers and glycopolymers are purified by continuous diafiltration using stirred ultrafiltration cells (Millipore) equipped with regenerated cellulose membranes (Ø 63.5 mm, Millipore) and connected to an auxiliary liquid reservoir pressurized with air; stirring rate ~300 rpm.
2. Glycomonomers: Initial concentration ≅ 50 g/L; washing through with 10 diafiltration volumes (DV) of 0.2 M $NaNO_3$ followed by 10 DV of deionized water, both at pH ≅ 6.5 (methacrylate monomers) or pH ≅ 8 (acrylamide and methacrylamide monomers).
3. Glycopolymers: Initial concentration ≅ 30 g/L; washing through 20 DV of 0.2 M $NaNO_3$, 0.01 M EDTA (pH 6), followed by 10 DV of deionized water, pH ≅ 6.5.
4. In both cases, purifications are stopped once the conductivity of the eluate had fallen below 5 μS/cm.

2.6 Degasification

1. Prior to polymerization, reaction mixtures are degassed and conditioned under N_2 atmosphere with a double vacuum/gas manifold connected to a rotary vane pump via a cold trap filled with liquid N_2, and to a purified N_2 gas line.

3 Methods

Two strategies are proposed for the synthesis of glycomonomers derived from negatively charged oligosaccharides (Fig. 1): (a) Acrylamide and methacrylamide monomers are prepared by reductive amination of the starting carbohydrate with NH_4AcO followed by acylation of the resulting 1-amino-1-deoxyalditol with (meth) acryloyl chloride [18], whereas (b) methacrylate and methacrylamide

Table 1
Reductive amination of glycuronan oligomers

Target molecule	Substrate, (mol L^{-1})	Amine (mol L^{-1})	$NaBH_3CN$ (mol L^{-1})	*t* (days)	*Q* (%)	Isolated yield (%)
$ManA_{11}$-NH_2	$ManA_{11}$, 0.028	NH_4OAc, 1.15	0.48	7	45	37
$GulA_{10}MA$	$GulA_{10}$, 0.017	AEM·HCl, 0.190	0.20	8	93	76

Conditions: water, stirring rate 250 rpm, T=20–30 °C. Conversion of the starting carbohydrate was >90 %; Q is the degree of functionalization of the recovered oligosaccharide

monomers are obtained in one step by reductive amination of the starting oligosaccharides with an excess of ethylenic monomer carrying a primary amino function [17]. Exact experimental conditions are summarized in Tables 1 and 2. After each step, salts and low molar mass species are eliminated by diafiltration, the degree of functionalization is determined by ^{1}H NMR and the obtained product is directly used for the following step (*see* also **Note 11**). The rationale behind this choice is that (1) the remaining non-aminated oligosaccharides are inert towards acylation (*see* **Note 1**) and (2) oligosaccharides not functionalized with an ethylenic group are inert towards radical polymerization (*see* **Note 12**). The oligosaccharides not incorporated into the final glycopolymer will be then eliminated by simple diafiltration (provided the hydrodynamic diameter of the two species is sufficiently different).

The obtained glycomonomers (*see* **Notes 13** and **14**) are copolymerized with HEMAm in the presence of 4,4′-azobis(cyanopentanoic acid) (ACPA) as radical initiator and 4-cyano-4-[(phenylcarbonothioyl) sulfanyl] pentanoic acid (CPCA) as the RAFT agent as shown in Fig. 2 (*see* **Note 15**). The reaction is carried out in a sodium acetate buffer in D_2O/DMSO-d6 in order to monitor its progress by ^{1}H NMR through the disappearance of the vinyl protons' peaks. Note that the acidity of the medium (pD = 5.2 ± 0.1) is essential to enable a controlled polymerization process [25]. Exact experimental conditions for polymerization are summarized in Table 3, whereas Table 4 recapitulates the physicochemical characteristics of the obtained polymers.

The fraction of dormant polymer chains in the glycopolymer (i.e., those bearing a trithiocarbonyl group at their ω-end) is estimated according to the method of Chieffari et al. [26]: Of the total number of polymer chains produced by a RAFT process, a number equal to those initiated by RAFT agent-derived radicals will possess a thiocarbonylthio end group and hence remain dormant, whereas a number equal to those initiated by primary (i.e., initiator derived) radicals will terminate irreversibly. It follows that the fraction of dormant chains in a sample (L) will be given by the ratio between the number of RAFT agent molecules and the number of chains produced by primary radicals *plus* the number of RAFT agent molecules:

Table 2
Synthesis of $ManA_{11}MAm$ by Method A

Substrate (mol L^{-1})	Buffer	Acylating agent (mol L^{-1})	Buffer, (mol L^{-1})	*Q* (%)	Isolated yield (%)
$ManA_{11}NH_2$, 0.011	Sodium carbonate, pH 9.5	Methacryloyl chloride, 0.15	0.60	45	99

Conditions: stirring rate 250 rpm, $T = 0\ °C \rightarrow 30\ °C$. *Q* is the degree of functionalization of the recovered oligosaccharide

Table 3
Summary of RAFT copolymerization experiments

	Glycomonomer							
Run	Type	mmol L^{-1}	HEMAm (mol L^{-1})	CPCA (mmol L^{-1})	ACPA (mmol L^{-1})	*t* (min)	*p* (%)	*T* (°C)
1	$ManA_{11}MAm$	36	0.56	5.6	1.9	430	95	60
2	$GulA_{10}MA$	16	0.48	2.7	1.3	420	89	70

Table 4
Characterization of obtained glycopolymers including the molar (*f*) and mass fraction (f_m) of glycomonomer in the feed, the control level $M_n/M_{n,th}$ achieved (*see* Note 16), and the fraction of dormant chains *L*

	Glycomonomer						
Glycopolymer	*f*	f_m	d*n*/d*c* (mL g^{-1})	M_n (Da)	*Đ*	$M_n/M_{n,th}$	*L* (%)
Poly(HEMAm-co-$ManA_{11}MAm$)	0.061	0.48	0.187	23,900	1.06	0.99	90
Poly(HEMAm-*co*-$GulA_{10}MA$)	0.031	0.36	0.192	35,900	1.09	1.23	70

$$L = \frac{[\text{RAFT}]_0}{[\text{RAFT}]_0 + f\left([\text{I}]_0 - [\text{I}]\right)(1+d)} \tag{2}$$

where $[\text{RAFT}]_0$ and $[\text{I}]_0$ are the initial concentrations of RAFT agent and initiator, respectively, $[\text{I}]$ is the final concentration of initiator, *f* is the initiator efficiency and *d* is the contribution of disproportionation to the overall termination process ($0 \le d \le 1$). In our case we used $d = 0.65$ and initiator decomposition rate constants $k_d = 1.19 \times 10^{-5}\ s^{-1}$ at 60 °C and $k_d = 4.85 \times 10^{-5}\ s^{-1}$ at 70 °C (*see* **Note 17**).

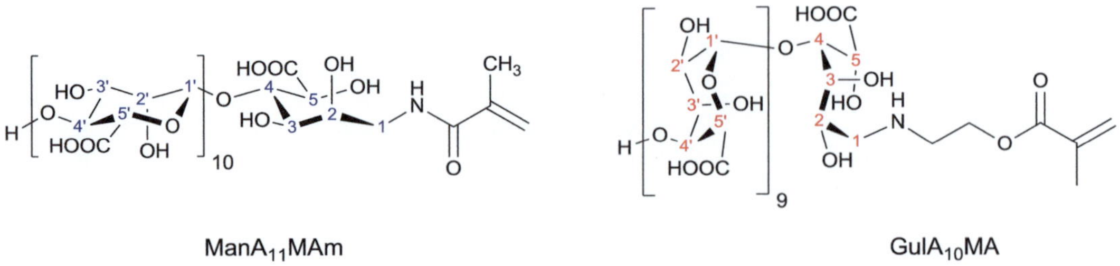

Fig. 3 Structure of the prepared glycomonomers and nuclei numbering used for NMR assignment

3.1 Preparation of Glycomonomers

The following protocols exemplify the transformation of (1→4)-β-D-mannuronan and (1→4)-α-L-guluronan into the vinyl glycomonomers shown in Fig. 3.

3.1.1 (1→4)-β-D-Mannuronan 1-Amino-1-deoxyalditol, $ManA_{11}NH_2$, Method A

1. Dissolve NH_4OAc (9.08 g, 115 mmol) in 35 mL of H_2O to make NH_4OAc solution.
2. Dissolve $NaBH_3CN$ (3.21 g, 48.0 mmol) in 15 mL of H_2O to make $NaBH_3CN$ solution.
3. Add 4.50 g of $ManA_{11}$ (2.79 mmol), 50 mL of H_2O, and the solution of NH_4OAc and $NaBH_3CN$ into a flask. Final pH≅7. The flask is then stoppered with a rubber septum and oxygen is removed by nitrogen sparging.
4. Stir the resulting solution at 200 rpm and 30 °C for 7 days.
5. Transfer the reaction mixture to a centrifugation tube and add ethanol under vigorous stirring to precipitate the oligosaccharide (final proportion of ethanol 80 % v/v).
6. Centrifuge at 12,000 *g* for 10 min.
7. Dissolve the precipitate in 80 mL of H_2O and diafilter it (NMWCO 500 Da, $p = 3.5$ bar).
8. Freeze-dry the purified solution to afford 4.07 g of white fluffy solid which is analysed by ^1H NMR. Conversion of starting carbohydrate, >90 %. Degree of functionalization (*Q*), 45 % (*see* **Notes 18** and **19**). Yield of $ManA_{11}NH_2$, 37 % (*see* **Note 20**).

3.1.2 (1→4)-β-D-Mannuronan Methacrylamide, $ManA_{11}MAm$, Method A

1. Dissolve 1.50 g (0.357 mmol) of the $ManA_{10}NH_2$ obtained in Subheading 3.1.1 in 34 mL of sodium carbonate buffer/MeOH (9:1) and cool the solution in an ice water bath for 10 min.
2. Add 493 μL (5.09 mmol) of methacryloyl chloride dropwise under vigorous stirring. From time to time adjust the pH to ~9.5 with solid Na_2CO_3.

3. After 2 h remove the ice water bath and stir the reaction mixture at T_{amb} for another 4.5 h.
4. Same work-up as in Subheading 3.1.1 affords 1.49 g of white fluffy solid. Conversion of $ManA_{10}NH_2$, 100 %. *Q*, 45 % (*see* **Note 18**). Yield, 99 % (*see* **Note 21**).

3.1.3 (1→4)-α-L-Guluronan Methacrylate, $GulA_{10}MA$, Method B

1. Charge the flask with 4.04 g of $GulA_{10}$ (1.93 mmol), 3.56 g of AEM·HCl (21.5 mmol, *see* **Note 8**), 0.508 g of $NaBH_3CN$ (7.68 mmol), 60 mL of acetate buffer (0.11 M, pH 5.9), and 54 mL of water.
2. Adjust the pH of the resulting solution to 5.3 with 1 M NaOH.
3. Stopper the flask with a rubber septum and remove oxygen by nitrogen sparging.
4. Stir the mixture at T_{amb} and 400 rpm.
5. Add $NaBH_3CN$ portions (0.50 g, 7.5 mmol) in the second and fourth day of reaction.
6. After 7 days, remove a brown precipitate by filtration on synthered glass filter and transfer the filtrate to a diafiltration cell for purification (NMWCO 500 Da, $p = 3.5$ bar).
7. Freeze-dry the purified solution to afford 3.5 g of white fluffy solid which is analysed by ^1H NMR (*see* **Note 22**). Conversion of starting oligosaccharide, 100 %. *Q*, 93 %. Yield of $GulA_{10}MA$, 76 %.

3.2 Preparation of Glycopolymers

3.2.1 Poly(HEMAm-co-$ManA_{11}MAm$) dithiobenzoate

1. Polymerization is carried out in an NMR tube equipped with a J. Young valve.
2. HEMAm, ACPA, and CPCA stock solutions of known concentration are used (*see* **Note 23**). Once prepared, these solutions must be kept at 4 °C in the dark until needed.
3. Dissolve 1.97×10^{-2} g (7.03×10^{-2} mmol) of the initiator in 1.00 mL of DMSO-d6, cool the resulting mixture to ~8 °C, and dilute it with an equal volume of deuterated acetate buffer to afford ACPA stock solution, $c_{ACPA} = 3.51 \times 10^{-2}$ M.
4. Dissolve 2.38×10^{-2} g (8.53×10^{-2} mmol) of the RAFT agent in 1.00 mL of DMSO-d6 to afford CPCA stock solution, $c_{CPCA} = 8.53 \times 10^{-2}$ M.
5. Dissolve 0.600 g (4.65 mmol) of monomer in 2.65 mL of deuterated acetate buffer and filter the resulting mixture through a 0.22 μm syringe filter (Nylon) to remove the suspended inhibitor (BHT) to afford HEMAm stock solution, $c_{HEMAm} = 1.75$ M.
6. In a wide-mouth glass vial, dissolve 0.059 g (*Q* 45 %, 3.20×10^{-3} mmol) of $ManA_{11}MAm$ in 500 μL of deuterated acetate buffer and mix the resulting solution with 280 μL

(0.491 mmol) of HEMAm, 57.6 μL (4.92×10^{-3} mmol) of CPADB and 47 μL (1.65×10^{-3} mmol) of ACPA stock solution.

7. Transfer the resulting mixture to the NMR tube which is sealed, degassed with three freeze-evacuate-thaw cycles, refilled with N_2 gas, and transferred to a water bath preheated at 60 °C.
8. After 428 min plunge the tube into ice water to quench the polymerization and transfer to an NMR spectrometer for ^{1}H NMR analysis (328 K, acquisition time = 2 s, relaxation delay = 14 s). Conversion 95 %.
9. Part of the reaction mixture is then sampled out for aqueous SEC-MALLS analysis: dn/dc 0.187, M_n 23,900, $Đ$ 1.06.

3.2.2 Poly(HEMAm-co-GulA$_{10}$MA)dithiobenzoate

1. Polymerization carried out in a 20 mL Schlenk tube equipped with a magnetic bar and stoppered with a silicone rubber septum.
2. HEMAm, ACPA, and CPCA stock solutions of known concentration are used (*see* **Note 23**). Once prepared, these solutions must be kept at 4 °C in the dark until needed.
3. Dissolve 2.50×10^{-2} g (8.74×10^{-3} mmol) of the initiator in 2.00 mL of D_2O with the help of a few grains of $NaHCO_3$ to afford ACPA stock solution, $c_{ACPA} = 4.37 \times 10^{-2}$ M, pD = 5.9.
4. Dissolve 1.81×10^{-2} g (6.48×10^{-3} mmol) of the RAFT agent in 2.00 mL of DMSO-d6 to afford CPCA stock solution, $c_{CPCA} = 3.24 \times 10^{-2}$ M.
5. Dissolve 4.66 g (3.61×10^{-2} mol) of monomer in 51.6 mL of D_2O then filter through two syringe filters connected in line (0.45 μm + 0.22 μm, Nylon) to remove the suspended inhibitor (BHT) to afford HEMAm stock solution, $c_{HEMAm} = 0.699$ M, 7.59 % w/w.
6. Add 0.250 g of GulA$_{10}$MA (Q 93 %, 0.113 mmol) to the Schlenk tube together with 5.50 g of the HEMAm stock solution (5.00 mL) and 1.40 mL of 1.0 M deuterated acetate buffer (pD 5.2).
7. Transfer the resulting mixture to an ice bath then add 0.600 mL (1.94×10^{-2} mmol) of CPCA and 0.220 mL of ACPA (9.62×10^{-3} mmol) stock solutions.
8. Seal the tube with a rubber septum, degass by four freeze-evacuate-thaw cycles, refill with N_2, and plunge it in an oil bath preheated at 70 °C.
9. Stir the reaction mixture for 7 h, and then quench it in an ice bath.
10. Draw 120 μL sample from the reaction tube and dilute it to 0.8 mL with D_2O for ^{1}H NMR analysis (328 K, acquisition time 4 s, relaxation delay 6 s). Conversion 89 %.

11. Dilute the remaining solution to 100 mL with of 0.2 M $NaNO_3$, 0.01 M EDTA (pH 6), and transfer it to a diafiltration cell for purification (NMWCO 10,000 Da, $p = 1.5$ atm.).
12. At the end of the process, freeze-dry the polymer in the dark. White-rose fluffy solid. Yield 538 mg, 86 % of the theoretical value. Aqueous SEC-MALLS analysis, dn/dc 0.192, M_n 35,900, $Đ$ 1.09.

4 Notes

1. In basic aqueous solution (Schotten–Baumann conditions) an acid chloride reacts rapidly with the primary amine of the 1-amino-1-deoxyalditol and any excess reagent is hydrolyzed by water ($c = 55$ M) before acylating the free hydroxyl groups of the carbohydrate ($c \leq 0.6$ M), the two reactions having similar rate constants [27].
2. HEMAm was chosen as a comonomer since it leads to highly hydrophilic copolymers [28] and because it closely resembles 2-hydroxypropylmethacrylamide (HPMAm) (the only difference being a methyl group on the side chain), whose homopolymer is non-toxic and non-immunogenic and whose copolymers have already been used for the preparation of bioconjugates for intravenous administration [29, 30]. Obviously, other hydrophilic methacrylamide derivatives can be used (*see* also **Note 4**).
3. RAFT polymerization of vinyl monomers combines the characteristics of a "living" polymerization process (i.e., homogeneous macromolecules, predetermined molar mass, dormant chain end) with the simplicity and robustness of radical polymerization [31]. It rests on the use of a thiocarbonylthio compound (e.g., a dithioester or trithiocarbonate) as the control agent (the "RAFT agent") and can be carried out under conditions that are important for glycopolymers' synthesis [32–36]: In homogeneous aqueous media [37, 38], at low-to-moderate temperature [39, 40, 32], and with monomers carrying complex functional groups (e.g., carboxylic acids) [41].
4. The lower limit for the molar mass of the oligosaccharides used is determined by the availability of a suitable purification method for the glycomonomers. Whereas dialysis is ineffective and time consuming, diafiltration is an appealing option when preparative aqueous size-exclusion chromatography is not available or the quantities involved exceed 100–200 mg. Although ultrafiltration membranes have a broad pore size distribution (hence are not highly selective) and no industry standard exists for determining their nominal molecular weight

cutoff (NMWCO), we found that negatively charged oligosaccharides with $M \geq$ NMWCO are well retained by Millipore Ultracel PL membranes (regenerated cellulose; eluent 0.2 M $NaNO_3$) in frontal flow diafiltration and by Pall Omega Membrane (polyethersulfone; eluent 0.1 M NaCl) in tangential flow diafiltration.

5. HEMAm: Viscous oil; colorless to pale yellow; TLC: petroleum ether/EtOAc/EtOH, 6:2:2, R_f 0.45; ^{1}H-NMR (δ, D_2O, 298 K): 1.93 (s, 3H, CH_3), 3.40 (t, 2H, J 5.7 Hz, NH-CH_2), 3.68 (t, 2H, J 5.7 Hz, CH_2-OH), 5.45 (s, 1H, CH_2=*cis* to CH_3), 5.71 (s, 1H, CH_2=*trans* to CH_3). ^{13}C-NMR (δ, D_2O, 283 K): 20.41 (CH_3), 44.34 (NH-CH_2), 62.60 (CH_2-OH), 123.95 (CH_2=), 141.61 (CH=), 174.83 (C=O).
6. After chromatographic purification, fractions containing HEMAm should be stabilized with BHT before being concentrated at the rotary evaporator and dried under mechanical vacuum. In the absence of a radical inhibitor, spontaneous polymerization is likely to occur.
7. CPCA: Amorphous solid; Red; TLC: petroleum ether/EtOAc/EtOH, 6:3:1, R_f 0.23; ^{1}H-NMR (δ, $CDCl_3$, 298 K): 1.94 (s, 3H, CH_3), 2.41–2.80 (m, 4H, CH_2-CH_2), 7.40 (m, 2H, meta-CH), 7.57 (m, 1H, para-CH), 7.91 (dd, 2H, *ortho*-CH, J 8.5 Hz, J 1.2 Hz). ^{13}C NMR (δ, $CDCl_3$, 298 K): 24.29 (CH_3), 29.68 (CH_2-COOH), 33.14 (CH_2-CH_2-COOH), 45.73 (C(CH_3)-C≡N), 118.51 (C≡N), 126.81 (*meta*-CH), 128.73 (*ortho*-CH), 133.22 (*CH*-C=S), 144.61 (*para*-CH), 177.35 (C=O), 222.30 (C=S).
8. The batch of 2-aminoethyl methacrylate hydrochloride used in this study (Aldrich ref. 516155, batch no. S47340-118) was over 96 mol% pure, according to ^{1}H NMR (*see* **Note 9**). Nevertheless, all other batches obtained from the same supplier contained up to 22 mol% of 2-aminoethyl 3-chloro-2-methylpropanoate and some residual copper (blue green colour; *see* **Note 10**).
9. 2-Aminoethyl methacrylate hydrochloride: ^{1}H NMR (δ, D_2O, 298 K): 1.94 (s, 3H, CH_3), 3.38 (t, 2H, J 5.1 Hz, CH_2-NH_2), 4.42 (t, 2H, J 5.1 Hz, CH_2-O), 5.77 (s, 1H, *CH2*=CH cis to CH_3), 6.19 (s, 1H, *CH2*=CH *trans* to CH_3). ^{13}C NMR (δ, D_2O, 298 K): 20.01 (CH_3), 41.21 (CH_2-NH_2), 63.96 (CH_2-O), 130.31 (*CH2*=CH), 137.95 (*CH*=CH_2), 171.75 (C=O).
10. 2-Aminoethyl 3-chloro-2-methylpropanoate: ^{1}H NMR (δ, D_2O, 298 K): 1.26 (d, 3H, J=7.1 Hz, CH_3), 3.15–3.03 (m, 1H, *CH*-CH_3), 3.79 (ddd, 2H, J=17.8, 11.0, 5.8 Hz, CH_2-Cl). ^{13}C NMR (δ, D_2O, 298 K): 16.71 (CH_3), 41.09 (CH_2-NH2), 44.49 (CH_2-Cl), 48.78 (*CH*-CH_3), 64.21 (CH_2-O), 178.30 (C=O). ESI-MS m/z, calculated 165.06

(100.0 %), 166.06 (6.7 %), 167.05 (32.0 %); found $[M+H]^+$ 166.3 (100.0 %), 167.3 (6.6 %), 168.3 (32.7 %).

11. Stoichiometric calculations have to take into account the degree of functionalization *Q* of the oligosaccharides.
12. Carbohydrates readily react with oxygen-centered radicals by hydrogen-atom abstraction at the ring C-H bonds [42] (for this reason peroxy initiators should not be used for the polymerization of vinyl glycomonomers) but are unaffected by most of the secondary and tertiary carbon-centered radicals encountered in radical polymerization processes (alkyl ester radicals being a notable exception) [43].
13. If the parent oligosaccharide is sufficiently bulky (i.e., at least five negatively charged monosaccharide units), the obtained glycomonomer will have little tendency to homopolymerize even in the presence of a radical initiator.
14. Glycomonomers derived from negatively charged oligosaccharides are best stored as a dry powder in their Na^+-salt form at 4 °C. For long-term storage (over 1 month), −18 °C is recommended.
15. In RAFT polymerization the presence of nucleophiles (e.g., free amines) and basic pH values should be avoided in order to prevent the lysis of the thiocarbylthio-compound (the RAFT agent) controlling the growth of macromolecules [44, 45, 20].
16. $M_{n,th}$ was calculated assuming a polymer composition identical to the feed (note that this hypothesis is strictly valid only for quantitative monomer conversion) and without taking into account initiator-derived chains. The following formula was used:

$$M_{\mathrm{n,th}} = p \cdot \frac{[\mathrm{M}_1]_0 + [\mathrm{M}_2]_0}{[\mathrm{CPCA}]_0} \cdot (f_1 M_1 + f_2 M_2) + M_{\mathrm{CPCA}} \qquad (3)$$

where M_1, M_2, and M_{CPCA} are the molecular masses of monomer 1 (i.e., HEMAm), monomer 2 (i.e., macromonomer) and the RAFT agent respectively; *p* is total monomer conversion from ^{1}H NMR, f_1 and f_2 are the molar fractions of monomer 1 and 2 in the feed, $[M_1]_0+[M_2]_0$ is the overall initial monomer concentration and $[RAFT]_0$ is the initial concentration of the RAFT agent. The exact $([M_1]_0+[M_2]_0)/[RAFT]_0$ ratios was determined by ^{1}H NMR analysis of the initial polymerization mixtures.

17. Initiator decomposition rate constants k_d were calculated from the kinetic parameters published by Overberger et al. [46] for the decomposition of *meso*-ACPA in DMAc.
18. The degree Rates of functionalization for the oligosaccharides was estimated by ^{1}H NMR: The peak of the internal anomeric proton H1′ (4.6 ppm in $ManA_{11}$ and 5.0 ppm in $GulA_{10}$);

being unaffected by the reaction, it was used as internal standard for integration. The integral of the newly formed *CH_2*-NH_2 signals (H1 at 3.0 and 3.4 ppm) or ethylenic protons signals (5.4 and 5.7 ppm for MAm; 5.8 and 6.2 ppm for MA) was compared to that of $H1_{\alpha/\beta}$ in the starting oligosaccharide.

19. Higher degrees of functionalization are obtained by adding $NaBH_3CN$ portion-wise every 24–48 h as in Subheading 3.1.3.
20. $ManA_{11}NH_2$, ^{1}H-NMR (δ, D_2O, 323 K): 3.00 (dd, 1H, $J_{1a,1b}$ 12.9 Hz, $J_{1a,2}$ 9.4 Hz, H1a), 3.47 (dd, 1H, $J_{1b,1a}$ 13.1 Hz, $J_{1b,2}$ 3.0 Hz, H1b), 3.61–4.30 (H2–H5, H2′–H5′), 4.62–4.81 (H1′), 5.02 (H1′, residual G unit).
21. $ManA_{11}MAm$, ^{1}H NMR (δ, D_2O, 323 K): 1.94 (m, 3H, CH_3), 3.38 (dd, 1H, $J_{1a,1b}$ 14.0 Hz, $J_{1a,2}$ 7.6 Hz, H1a), 3.59–4.34 (H2–H5, H2′–H5), 4.64–4.80 (H1′), 5.00 (H1′, residual G unit), 5.44 (m, 1H, *CH_2*=CH *cis* to CH_3), 5.70 (m, 1H, *CH_2*=CH *trans* to CH_3).
22. $GulA_{10}MA$: ^{1}H-NMR (δ, D_2O, 308 K): 1.94 (s, 3H, CH_3,), 3.20 (dd, 1H, $J_{1a\text{-}1b}$ 12.8 Hz, $J_{1a\text{-}2}$ 9.6 Hz, H1a), 3.45–3.5 (br, 3H, H1b and CH_2-CH_2-NH), 3.89 (m, H2, H2′), 4.00 (m, H4, H4′), 4.12 (m, H5, H5′), 5.00–5.17 (m, H1′), 5.77 (s, 1H, *CH_2*=CH *cis* to CH_3), 6.18 (s, 1H, *CH_2*=CH *trans* to CH_3).
23. Stock solutions of reagents are best prepared in wide-mouth glass vials with a screw cap as to allow safe storage into a refrigerator and liquid handling with an automatic pipette.

Acknowledgements

This work was supported by the *Cluster de Recherche Chimie Durable et Chimie pour la Santé* of the Rhône-Alpes region, the competitiveness cluster Axelera (Lyon, France), and the *Agence Nationale de la Recherche* (ANR-09-CP2D-02 ALGIMAT).

References

1. Sigal GB, Mammen M, Dahmann G, Whitesides GM (1996) Polyacrylamides bearing pendant alpha-sialoside groups strongly inhibit agglutination of erythrocytes by influenza virus—the strong inhibition reflects enhanced binding through cooperative polyvalent interactions. J Am Chem Soc 118(16):3789–3800
2. Kanai M, Mortell KH, Kiessling LL (1997) Varying the size of multivalent ligands—the dependence of concanavalin A binding on neoglycopolymer length. J Am Chem Soc 119(41):9931–9932
3. Kiessling LL, Gestwicki JE, Strong LE (2000) Synthetic multivalent ligands in the exploration of cell-surface interactions. Curr Opin Chem Biol 4(6):696–703
4. Ghadban A, Albertin L (2013) Synthesis of glycopolymer architectures by reversible-deactivation radical polymerization. Polymers 5(2):431–526
5. Spain SG, Gibson MI, Cameron NR (2007) Recent advances in the synthesis of well-defined glycopolymers. J Polym Sci A Polym Chem 45(11):2059–2072

6. Narumi A, Matsuda T, Kaga H, Satoh T, Kakuchi T (2001) Glycoconjugated polymer II. Synthesis of polystyrene-*block*-poly(4-vinylbenzyl glucoside) and polystyrene-*block*-poly(4-vinylbenzyl maltohexaoside) via 2,2,6,6-tetramethylpiperidine-1-oxyl-mediated living radical polymerization. Polym J (Tokyo, Jpn) 33(12):939–945
7. Narumi A, Otsuka I, Matsuda T, Miura Y, Satoh T, Kaneko N, Kaga H, Kakuchi T (2006) Glycoconjugated polymer: synthesis and characterization of poly(vinyl saccharide)-*block*-polystyrene-block-poly(vinyl saccharide) as an amphiphilic ABA triblock copolymer. J Polym Sci A Polym Chem 44(13): 3978–3985
8. Sun XL, Grande D, Baskaran S, Hanson SR, Chaikof EL (2002) Glycosaminoglycan mimetic biomaterials. 4. Synthesis of sulfated lactose-based glycopolymers that exhibit anticoagulant activity. Biomacromolecules 3(5):1065–1070
9. Baskaran S, Grande D, Sun XL, Yayon A, Chaikof EL (2002) Glycosaminoglycan-mimetic biomaterials. 3. Glycopolymers prepared from alkene-derivatized mono- and disaccharide-based glycomonomers. Bioconjug Chem 13(6):1309–1313
10. Grande D, Baskaran S, Chaikof EL (2001) Glycosaminoglycan mimetic biomaterials. 2. Alkene- and acrylate-derivatized glycopolymers via cyanoxyl-mediated free-radical polymerization. Macromolecules (Washington, DC) 34(6):1640–1646
11. Grande D, Baskaran S, Baskaran C, Gnanou Y, Chaikof EL (2000) Glycosaminoglycan-mimetic biomaterials. 1. Nonsulfated and sulfated glycopolymers by cyanoxyl-mediated free-radical polymerization. Macromolecules (Washington, DC) 33(4):1123–1125
12. Guan R, Sun XL, Hou S, Wu P, Chaikof EL (2004) A glycopolymer chaperone for fibroblast growth factor-2. Bioconjug Chem 15(1):145–151
13. Kitano H, Saito D, Kamada T, Gemmei-Ide M (2012) Binding of β-amyloid to sulfated sugar residues in a polymer brush. Colloids Surf B 93:219–225
14. Varki A, Cummings R, Esko J, Freeze H, Hart G, Marth J (eds) (2009) Essentials of glycobiology, 2nd edn. Cold Spring Harbor Laboratory Press, Cold Spring Harbor, NY
15. Pawar SN, Edgar KJ (2011) Chemical modifications of alginates in organic solvent systems. Biomacromolecules 12 (11):4095–4103. doi:10.1021/bm201152a
16. van den Bos LJ, Codée JDC, Litjens REJN, Dinkelaar J, Overkleeft HS, van der Marel G,A (2007) Uronic acids in oligosaccharide synthesis. Eur J Org Chem 24:3963–3976
17. Ghadban A, Reynaud E, Rinaudo M, Albertin L (2013) RAFT copolymerization of alginate-derived macromonomers—synthesis of a well-defined poly(HEMAm)-graft-(1 → 4)-α-L-guluronan copolymer capable of ionotropic gelation. Polym Chem 4(17):4578–4583. doi:10.1039/c3py00730h
18. Ghadban A, Albertin L, Rinaudo M, Heyraud A (2012) Biohybrid glycopolymer capable of ionotropic gelation. Biomacromolecules 13(10):3108–3119
19. Moad G, Rizzardo E, Thang SH (2005) Living radical polymerization by the RAFT process. Aust J Chem 58(6):379–410
20. Moad G, Rizzardo E, Thang SH (2009) Living radical polymerization by the RAFT process—a second update. Aust J Chem 62(11):1402–1472
21. Spain SG, Albertin L, Cameron NR (2006) Facile in situ preparation of biologically active multivalent glyconanoparticles. Chem Commun (Cambridge, UK) (40):4198–4200
22. Thang SH, Chong YK, Mayadunne RTA, Moad G, Rizzardo E (1999) A novel synthesis of functional dithioesters, dithiocarbamates, xanthates and trithiocarbonates. Tetrahedron Lett 40(12):2435–2438
23. Huglin MB (1972) Specific refractive index increment. In: Huglin MB (ed) Light scattering from polymer solutions. Physical chemistry. Academic, London, pp 165–331
24. Rinaudo M (2007) Seaweed polysaccharides. In: Kamerling JP (ed) Comprehensive glycoscience, vol 2. Elsevier, New York, pp 691–735
25. Scales CW, Vasilieva YA, Convertine AJ, Lowe AB, McCormick CL (2005) Direct, controlled synthesis of the nonimmunogenic, hydrophilic polymer, poly(N-(2-hydroxypropyl)methacrylamide) via RAFT in aqueous media. Biomacromolecules 6(4):1846–1850
26. Chiefari J, Rizzardo E (2002) Control of free radical polymerization by chain transfer methods. In: Matyjaszewski K, Davis TP (eds) Handbook of radical polymerization. Wiley, Hoboken, NJ, pp 629–690
27. Bentley TW, Llewellyn G, McAlister JA (1996) SN2 mechanism for alcoholysis, aminolysis, and hydrolysis of acetyl chloride. J Org Chem 61(22):7927–7932. doi:10.1021/jo9609844
28. Rijcken CJF, Veldhuis TFJ, Ramzi A, Meeldijk JD, van Nostrum CF, Hennink WE (2005) Novel fast degradable thermosensitive polymeric micelles based on PEG-block-poly

(N-(2-hydroxyethyl)methacrylamide-oligo-lactates). Biomacromolecules 6(4):2343–2351

29. Duncan R (2009) Development of HPMA copolymer-anticancer conjugates: clinical experience and lessons learnt. Adv Drug Deliv Rev 61(13):1131–1148. doi:10.1016/j.addr.2009.05.007
30. Vicent MJ, Ringsdorf H, Duncan R (2009) Polymer therapeutics: clinical applications and challenges for development. Adv Drug Deliv Rev 61(13):1117–1120. doi:10.1016/j.addr.2009.08.001
31. Matyjaszewski K, Davis TP (eds) (2002) Handbook of radical polymerization. Wiley, Hoboken, NJ
32. Albertin L, Wolnik A, Ghadban A, Dubreuil F (2012) Aqueous RAFT polymerization of N-acryloylmorpholine, synthesis of an ABA triblock glycopolymer and study of its self-association behavior. Macromol Chem Phys 213(17):1768–1782
33. Albertin L, Cameron NR (2007) RAFT polymerization of methyl 6-*O*-methacryloyl-α-D-glucoside in homogeneous aqueous medium. A detailed kinetic study at the low molecular weight limit of the process. Macromolecules (Washington, DC) 40(17):6082–6093
34. Albertin L, Stenzel MH, Barner-Kowollik C, Foster LJR, Davis TP (2005) Well-defined diblock glycopolymers from RAFT polymerization in homogeneous aqueous medium. Macromolecules (Washington, DC) 38(22):9075–9084
35. Albertin L, Stenzel M, Barner-Kowollik C, Foster LJR, Davis TP (2004) Well-defined glycopolymers from RAFT polymerization: poly(methyl 6-*O*-methacryloyl-α-D-glucoside) and its block copolymer with 2-hydroxyethyl methacrylate. Macromolecules (Washington, DC) 37(20):7530–7537
36. Albertin L, Kohlert C, Stenzel M, Foster LJR, Davis TP (2004) Chemoenzymatic synthesis of narrow-polydispersity glycopolymers: poly(6-O-vinyladipoyl-D-glucopyranose). Biomacromolecules 5(2):255–260
37. Lowe AB, McCormick CL (2007) Reversible addition-fragmentation chain transfer (RAFT) radical polymerization and the synthesis of water-soluble (co)polymers under homogeneous conditions in organic and aqueous media. Prog Polym Sci 32(3):283–351
38. McCormick CL, Lowe AB (2004) Aqueous RAFT polymerization: recent developments in synthesis of functional water-soluble (co)polymers with controlled structures. Acc Chem Res 37(5):312–325
39. Quinn JF, Rizzardo E, Davis TP (2001) Ambient temperature reversible addition-fragmentation chain transfer polymerisation. Chem Commun (Cambridge, UK) (11): 1044–1045
40. Convertine AJ, Ayres N, Scales CW, Lowe AB, McCormick CL (2004) Facile, controlled, room-temperature RAFT polymerization of N-isopropylacrylamide. Biomacromolecules 5(4):1177–1180
41. Chaduc I, Lansalot M, D'Agosto F, Charleux B (2012) RAFT polymerization of methacrylic acid in water. Macromolecules (Washington, DC) 45(3):1241–1247. doi:10.1021/ma2023815
42. Hawkins CL, Davies MJ (1996) Direct detection and identification of radicals generated during the hydroxyl radical-induced degradation of hyaluronic acid and related materials. Free Radic Biol Med 21(3):275–290
43. Albertin L, Stenzel MH, Barner-Kowollik C, Foster LJR, Davis TP (2005) Solvent and oxygen effects on the free radical polymerization of 6-*O*-vinyladipoyl-D-glucopyranose. Polymer 46(9):2831–2835
44. Thomas DB, Convertine AJ, Hester RD, Lowe AB, McCormick CL (2004) Hydrolytic susceptibility of dithioester chain transfer agents and implications in aqueous RAFT polymerizations. Macromolecules (Washington, DC) 37(5):1735–1741
45. Albertin L, Stenzel MH, Barner-Kowollik C, Davis TP (2006) Effect of an added base on (4-cyanopentanoic acid)-4-dithiobenzoate mediated RAFT polymerization in water. Polymer 47(4):1011–1019
46. Overberger CG, Labianca DA (1970) Azo compounds. 48. Optically active azonitriles. J Org Chem 35(6):1762–1770

Chapter 3

Glycopolymers Prepared by Ring-Opening Metathesis Polymerization Followed by Glycoconjugation Using a Triazole-Forming "Click" Reaction

Ronald Okoth and Amit Basu

Abstract

We describe a protocol for the preparation of glycopolymers derived from the ring-opening polymerization of a norbornene carboxylic acid derivative. Polymerization is followed by attachment of a linker and subsequent glycoconjugation via a triazole-forming azide-alkyne click reaction. The use of a protected amine-terminating agent allows for the attachment of a probe molecule such as a fluorescein dye. The syntheses of a neutral galactopolymer as well a polyanionic poly-3-*O*-sulfo-galactopolymer are described.

Key words Glycopolymer, Ring-opening metathesis polymerization, Click chemistry, Triazole, Fluorescent glycoconjugate, Linker, Propargyl glycoside, Galactose, Sulfo-galactose

1 Introduction

Polymers functionalized with carbohydrate-bearing side chains, or glycopolymers, are important tools for studying carbohydrate recognition phenomena [1]. They can be prepared in a variety of lengths and can readily be functionalized with reporter molecules such as dyes that aid in visualizing and quantifying binding to cell surface receptors and chemically modified surfaces. Polymers prepared via ring-opening metathesis polymerization (ROMP) have proven to be particularly effective as multivalent glycoconjugates, with applications ranging from probing bacterial signaling [2] to chemokine-glycosaminoglycan interactions [3].

We have recently reported a highly modular approach for the synthesis of glycopolymers that relies on the use of the copper promoted azide-alkyne coupling to covalently attach carbohydrates to the polymer backbone (Fig. 1) [4]. Additionally, the use of a Teoc (2-trimethylsilylethyl carbamate) protected-amine terminating agent (*TA*) provides a useful NMR handle for determining the degree of polymerization and the success of subsequent

Xue-Long Sun (ed.), *Macro-Glycoligands: Methods and Protocols*, Methods in Molecular Biology, vol. 1367,
DOI 10.1007/978-1-4939-3130-9_3, © Springer Science+Business Media New York 2016

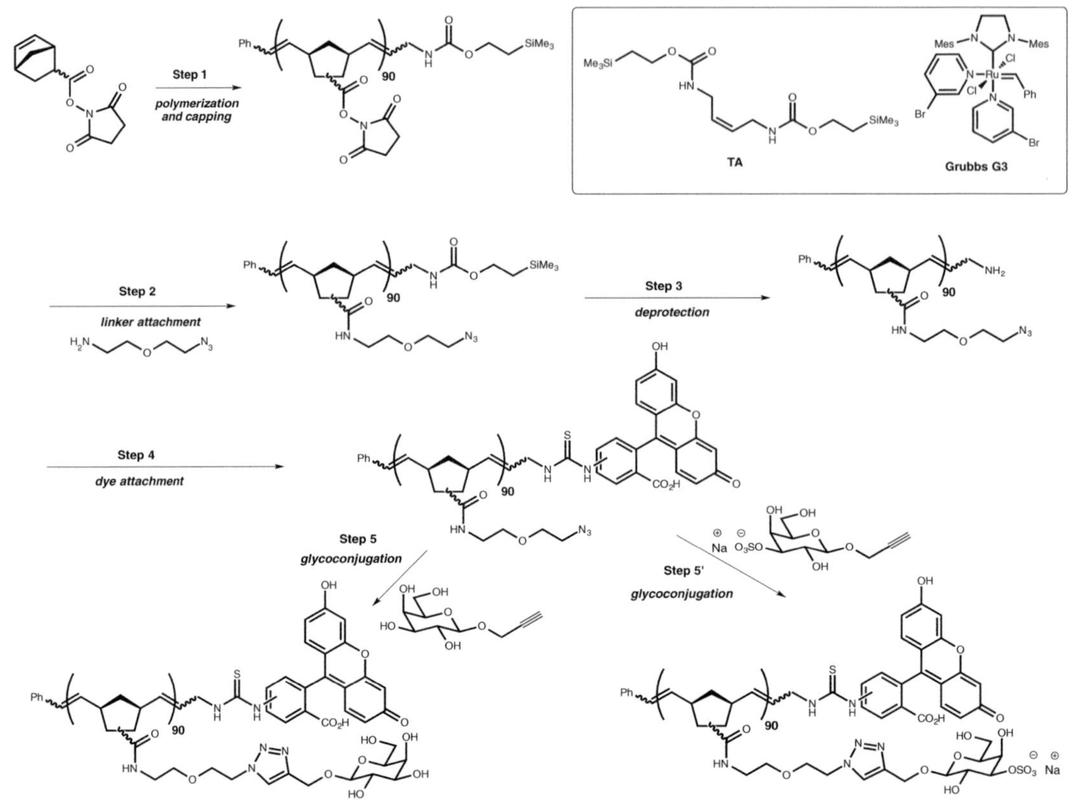

Fig. 1 Synthetic scheme for polymerization, labeling, and glycoconjugation to provide dye-labeled ROMP glycopolymers

post-polymerization modifications. Upon deprotection, the amino group provides a site for further conjugation, such as dye attachment.

The synthesis of fluorescein-terminated ROMP glycopolymers is described. Polymerization of an activated ester of norbornene carboxylic acid is followed by termination using a carbamate-protected amine. Amidation of the activated ester is carried out using an azide-terminated linker. Deprotection of the carbamate and subsequent attachment of the dye provides the penultimate intermediate, which can be functionalized with the desired alkyne containing sugar. Copper-promoted triazole formation using propargyl glycosides provides the target glycoconjugates.

2 Materials

2.1 Chemicals

1. Grubbs' generation three catalyst (*Grubbs G3*) is prepared as reported by Love et al. [5].
2. Propyl β-D galactopyranoside is prepared as reported by Zhao et al. [6].

3. 3′-*O*-sulfo-1-propyl β-D galactopyranoside is prepared as reported by Zhao et al. [6].
4. 2-(2-(2-Aminoethoxy)-ethoxy)ethanol tris(3-hydroxypropyl triazolylmethyl)amine (THPTA) ligand is prepared as reported by Hong et al. [7].
5. The terminating agent and NHS ester is synthesized as reported by Okoth and Basu [4].
6. The linker 2-(2-azidoethoxy)-ethan-1-amine is prepared as reported by Chouhan and James [8].

2.2 Other Reagents

1. Ethyl vinyl ether is obtained from Sigma-Aldrich.
2. Tetrabutylammonium fluoride solution in THF is obtained from Sigma-Aldrich.
3. Fluorescein isothiocyanate is obtained from Sigma-Aldrich.
4. Sodium ascorbate is obtained from Sigma-Aldrich.
5. Copper sulfate is obtained from Fisher Scientific.
6. Dry solvents are obtained from a commercially available solvent purification system based on protocols reported by Pangborn et al. [9].

2.3 Instrumentation

7. Dialysis is performed using Spectra/Por 6-regenerated cellulose dialysis membrane 1000 MWCO from Spectrum.
8. Concentration by centrifugal filtration is done using Amicon Ultra centrifugal filters (regenerated cellulose 3000 MWCO) filter devices in a Sorvall ST 16 centrifuge from Thermo Scientific operated at 4000 × *g*.
9. NMR spectra are recorded on Bruker 600, 400, or 300 MHz instruments for ^{1}H NMR, and are referenced at 2.05 ppm for d_6-acetone and 4.79 ppm for D_2O. ^{1}H NMR spectra for all polymer products are obtained with a delay time of 5 s.
10. An Agilent 8453 UV-visible spectroscopy system is used to obtain UV-vis spectra of end-labeled glycopolymer solutions in a 1 cm quartz cell.
11. Microwave reactions are carried out in a Biotage Initiator Microwave Synthesizer in 2–5 mL microwave reaction vials.

3 Methods

The first step involves ROMP of the *N*-hydroxysuccinimide (NHS) ester of norbornene carboxylic acid using Grubbs' third-generation catalyst (Grubbs' G3). The polymerization initiates rapidly, so the reaction is set up under nitrogen at −78 °C and subsequently removed from the dry ice-acetone bath and allowed to warm to room temperature. The polymerization reaction is complete within 30–40 min (as judged by TLC—silica gel/CH_2Cl_2). The polymer is then capped using the terminating agent (*TA*) derived from *cis-1*,4-diamino-2-butene. Termination occurs quantitatively within 20 min using a fivefold excess of the TA. The Teoc-protecting group is useful for two reasons—(1) the nine protons on the TMS group provide a strong NMR signal that can be used to determine the degree of polymerization; and (2) removal of the protecting group provides an amino functionality that can be used for subsequent conjugation chemistry (**step 4**).

The second step involves reaction of the pendant NHS ester moieties on the ROMP-generated polymer obtained in **step 1** with an α-ω-azido-amine linker. While glycoconjugation can be carried out directly on the NHS ester moieties via amidation, the installation of the pendant azide groups achieved in this step allows for the use of the high yielding copper promoted click reaction, which can be carried out with complex functionalities in aqueous solution.

Our initial report on these glycopolymers described glycoconjugation in **step 3** followed by attachment of the dye [4]. While this approach does generate functionalized glycopolymer probes, it results in differences in the amounts of dye attached to polymers bearing different sugars as a result of batch-to-batch variation in the dye-labeling step. The current protocol desrcibes an alternative and more effective approach involving initial dye attachment followed by glycoconjugation. This results in uniform labeling of multiple glycopolymers derived from a single batch of dye-labeled polymer. Thus, **step 3** consists of Teoc group deprotection using TBAF at 50 °C. Confirmation of removal of the Teoc group is obtained by ^{1}H NMR, which shows the disappearance of the resonance derived from the TMS protons.

The deprotection of the terminal amine provides an avenue for attaching diverse probes to the polymer using a variety of chemistries. We describe here the use of fluorescein isothiocyanate (FITC), resulting in the formation of a thiourea-linked dye. As with the previous deprotection step, purification of the polymer is carried out by extensive dialysis.

The final step, glycoconjugation, is carried out via a click reaction using copper sulfate and sodium ascorbate as promoters for the coupling between the pendant azide groups and propargyl glycosides. The use of a microwave permits the reaction to be completed in 10 min, although functionalization with an anionic sulfo-sugar takes considerably longer. The success of the click reaction can be readily monitored using IR spectroscopy by tracking the disappearance of the strong azide band at 2100 cm^{-1}.

3.1 Synthesis of Polynorbornyl NHS Ester (Step 1)

1. Add the NHS ester of norbornene carboxylic acid (1.4 g, 5.95 mmol) directly into a 20 mL scintillation vial, followed by 10 mL of dry CH_2Cl_2 and a magnetic stirring bar.
2. Cap the vial with a rubber septum, place the solution under a gentle stream of nitrogen using an inlet needle (*see* **Note 1**) and then cool it to −78 °C in a dry ice-acetone bath atop a stirring plate (*see* **Note 2**).
3. Add a solution of Grubbs' generation 3 (Grubbs G3) catalyst (0.06 g, 0.066 mmol) in 5 mL of dry CH_2Cl_2 to the vial quickly via a syringe while stirring the mixture (*see* **Note 3**).
4. Remove the dry ice-acetone bath carefully and allow the reaction to warm to room temperature (*see* **Note 4**).
5. After 40 min add a solution of *TA* (0.124 g, 0.33 mmol, five equivalents relative to catalyst) in 5 mL of dry CH_2Cl_2 to the reaction via a syringe.
6. After another 20 min add 0.5 mL of ethyl vinyl ether (0.38 g, 5.22 mmol) via a syringe to quench the reaction.
7. Transfer the reaction solution using a Pasteur pipette to four different test tubes, each containing about 10 mL of ether, forming a light brown precipitate.
8. Centrifuge the test tubes and then decant the ether (*see* **Note 5**).
9. Wash the precipitate by adding 10 mL of ether into the test tubes and agitating the mixture with a Pasteur pipette. Centrifuge the test tubes and then decant the ether again. Repeat the process of centrifuging and decanting until the ether is clear.
10. Collect the grey precipitate and dry under vacuum to give the polynorbornyl NHS ester (1.09 g, 0.05 mmol) as a grey solid in 77 % yield (*see* **Notes 6** and 7).

3.2 Synthesis of Polynorbornyl Azide (Step 2)

1. Dissolve the polynorbornyl NHS ester (0.6 g, 0.03 mmol) in 10 mL of dry CH_2Cl_2 in a round-bottom flask.
2. Dissolve the α-ω-azido-amine linker (0.7 g, 5.4 mmol, two equivalents for each NHS moiety on the polymer) in 5 mL dry CH_2Cl_2 in a vial, transfer it to the flask above, and stir the solution for 24 h at ambient temperature.
3. Remove the solvent using a rotary evaporator.
4. Dissolve the resulting brown residue in 10 mL of MeOH then transfer it into a dialysis tube and dialyze in 200 mL of MeOH reservoir. Change the solvent every 12 h over 36 h.
5. Collect the solution in the dialysis tube and evaporate it on a rotary evaporator. Dry the product under vacuum to provide

the polynorbornyl azide as a light brown sticky gel (0.5 g, 0.022 mmol) in 73 % yield (*see* **Note 8**).

3.3 Deprotection of Polynorbornyl Azide (Step 3)

1. Dissolve the polynorbornyl azide (*see* **Note 9**) (0.4 g, 0.018 mmol) in 5 mL of dry THF.
2. To this solution add 1 mL of TBAF in THF (ca. 1 mol/L) and heat the mixture at 50 °C overnight.
3. Remove the solvent using a rotary evaporator.
4. Dissolve the brown residue in 10 mL of MeOH, transfer it into a dialysis tube, and dialyze in 200 mL of DCM:MeOH (1:1,v/v) reservoir. Change the solvent reservoir every 12 h for 36 h.
5. Collect the solution in the dialysis tube, evaporate and dry under vacuum to give the deprotected polynorbornyl azide (0.35 g, 0.015 mmol) 83 % as a light brown sticky gel (*see* **Note 10**).

3.4 Synthesis of Fluorescein-Terminated Polynorbornyl Azide (Step 4)

1. Dissolve the deprotected polynorbornyl azide (0.33 g, 0.014 mmol) in 10 mL of DMF in a scintillation vial.
2. Add a solution of fluorescein isothiocyanate (FITC) in 2 mL of DMF (0.04 g, 0.1 mmol), followed by DMAP (0.002 g, 0.014 mmol) to the vial.
3. Cover the vial with aluminum foil and stir the reaction at 50 °C for 24 h.
4. Transfer the solution into a dialysis tube and dialyze against 400 mL of methanol reservoir until the methanol reservoir is clear (*see* **Notes 11** and **12**).
5. Collect the solution in the tube and evaporate it on a rotary evaporator to provide the fluorescein-terminated polynorbornyl azide (0.3 g, 0.013 mmol, 93 % yield) as a yellow film (*see* **Note 13**).

3.5 Synthesis of Galactose-Functionalized Glycopolymer (Step 5)

1. Sequentially charge a 2–5 mL microwave vial with a stir bar, followed by solutions of fluorescein-terminated polynorbornyl azide (0.1 g, 0.004 mmol) in 2 mL of MeOH, 1-propynyl β-D galactopyranoside (0.2 g, 0.92 mmol) in 0.5 mL of H_2O, $CuSO_4 \cdot 5H_2O$ (8 mg, 0.03 mmol) in 0.5 mL of H_2O, and Na-ascorbate (12 mg, 0.06 mmol) in 0.5 mL of H_2O.
2. Place the vial in a microwave and irradiate at 80 °C for 10 min.
3. Transfer the resulting solution into a vial containing 1 g of CupriSorb® [7] and 5 mL of H_2O. Stir the solution at ambient temperature overnight (*see* **Note 14**).
4. Filter the mixture through a cotton plug and wash the CupriSorb® with 10 mL of H_2O.

5. Dialyze the collected filtrate against 200 mL of H_2O. Change the solvent reservoir every 12 h for 48 h (*see* **Note 12**).
6. Collect the dialyzed solution and concentrate it to about 2 mL by centrifugal filtration using an Amicon filter.
7. Add toluene (1 mL) to the concentrated solution and evaporate the mixture on a rotary evaporator with the water bath maintained at 40 °C. Repeat this procedure two more times.
8. Dry the product under high vacuum to provide the galactose-functionalized glycopolymer (0.09 g, 2 μmol) as a brown film (see **Notes 15–17**).

3.6 Synthesis of 3′-O-Sulfo-galactose-Functionalized Glycopolymer (Step 5′)

1. Sequentially charge a scintillation vial with a stir bar, followed by solutions of fluorescein-terminated polynorbornyl azide (0.08 g, 0.003 mmol) in 5 mL of MeOH, 3′-*O* sulfo-1-propyl β-D galactopyranoside (0.2 g, 0.6 mmol) in 0.5 mL of H_2O, $CuSO_4 \cdot 5H_2O$ (8 mg, 0.03 mmol) and THPTA (65 mg, 0.15 mmol) mixture in 0.5 mL of H_2O, and Na-ascorbate (12 mg, 0.06 mmol) in 0.5 mL of H_2O.
2. Heat the capped scintillation vial at 50 °C for 24 h.
3. Add 10 mL of H_2O to the reaction solution, followed by 1 g of CupriSorb®. Stir the mixture at ambient temperature overnight.
4. Filter the solution through a cotton plug and wash the CupriSorb® with 10 mL of H_2O.
5. Dialyze the collected filtrate against 200 mL of H_2O. Change the solvent reservoir every 12 h for 48 h (*see* **Note 12**).
6. Collect the solution and concentrate it to about 2 mL by centrifugal filtration and transfer the solution to a scintillation vial.
7. Add toluene (1 mL) to the concentrated solution and evaporate the solution on a rotary evaporator with the water bath maintained at 40 °C. Repeat this procedure two more times.
8. Finally, add heptane (1 mL) and evaporate the solution on a rotary evaporator.
9. Dry the product under vacuum to provide the 3′-*O-sulfo-galactose-functionalized* glycopolymer (0.1 g, 2 μmol) as a light brown film (*see* **Note 18**).

4 Notes

1. The inlet needle should be derived from a manifold with a bubbler, or a vent needle should also be inserted into the vial.
2. The NHS ester is used as an *exo/endo* mixture, prepared from the commercially available *exo/endo* norbornene carboxylic acid.

3. If catalyst remains in the syringe, another 1 mL of dry CH_2Cl_2 may be drawn up into the syringe and added to the reaction vial to rinse any residual catalyst out of the syringe.
4. Upon warming, the solution turns from bright green to brown after about 2 min.
5. Centrifugation was done using an Adams physician's compact centrifuge.
6. The % yield is based on the ideal ($n = 90$) polymer theoretical molecular weight (21,434 g).
7. Partial ^{1}H NMR (400 MHz, d_6-acetone) δ 7.33 (m, 5H, *Ph*) 5.76–5.37 (m, 194H, *olefinic-H*) 0.04 (s, 9H, *Me3Si*).
8. Confirmation of the substitution was obtained by the appearance of a strong azide band in the IR spectrum at 2100 cm^{-1}. Disappearance of the ^{1}H NMR peaks from the NHS moiety could not be used to quantitatively assess the degree of amidation because of signal overlap.
9. The polynorbornyl azide, as well as many subsequent derivatives, are sticky and cannot easily be weighed out. In these cases a stock solution is made and the appropriate amount for the reaction is dispensed volumetrically.
10. The disappearance of the TMS peaks at 0 ppm provided confirmation of complete deprotection.
11. Completion of dialysis took 5 days, with the reservoir exchanged every 12 h.
12. To minimize exposure to light this setup is covered with aluminum foil.
13. The amount of dye loading was quantified after glycoconjugation, *see* **Note 16**.
14. Cuprisorb® is a copper chelating resin that is readily available in pet stores, as it is used in fish tanks.
15. Partial ^{1}H NMR (600 MHz, D_2O) δ 8.05 (triazole H), 5.26 (olefinic H), 4.55–3.50, sugar H.
16. Dye loading was quantified by measuring the absorbance of the polymer (2–10 μM, @ 497 nm) and comparing these values against a calibration curve generated with FITC. Loading was found to be quantitative.
17. The final glycopolymers are dissolved in deionized water (@ 0.5 M) and stored in a scintillation vial. The vial is covered with foil, wrapped tightly with parafilm and stored in the fridge. They can be stored for up to 2–3 months in this form. When needed for assays, the vials are allowed to come to room temperature, and the stock solution is diluted in buffer.
18. Partial ^{1}H NMR (600 MHz, D_2O) δ 7.99 (triazole H), 5.15 (olefinic H), 4.88–3.78 (sugar H).

Acknowledgement

This work was supported by the National Science Foundation (CHE-0910947).

References

1. Kiessling LL, Grim JC (2013) Glycopolymer probes of signal transduction. Chem Soc Rev 42:4476–4491
2. Gestwicki JE, Strong LE, Kiessling LL (2000) Tuning chemotactic responses with synthetic multivalent ligands. Chem Biol 7:583–591
3. Sheng GJ, Oh YI, Chang S-K, Hsieh-Wilson LC (2013) Tunable heparan sulfate mimetics for modulating chemokine activity. J Am Chem Soc 135:10898–10901
4. Okoth R, Basu A (2013) End-labeled amino terminated monotelechelic glycopolymers generated by ROMP and Cu(I)-catalyzed azide–alkyne cycloaddition. Beilstein J Org Chem 9:608–612
5. Love JA, Morgan JP, Trnka TM, Grubbs RH (2002) A practical and highly active ruthenium-based catalyst that effects the cross metathesis of acrylonitrile. Angew Chem Int Edit 41:4035–4037
6. Zhao J, Liu Y, Park HJ, Boggs JM, Basu A (2012) Carbohydrate-coated fluorescent silica nanoparticles as probes for the galactose/3-sulfogalactose carbohydrate–carbohydrate interaction using model systems and cellular binding studies. Bioconjugate Chem 23:1166–1173
7. Hong V, Presolski SI, Ma C, Finn MG (2009) Analysis and optimization of copper-catalyzed azide-alkyne cycloaddition for bioconjugation. Angew Chem Int Edit 48:9879–9883
8. Chouhan G, James K (2011) CuAAC macrocyclization: high intramolecular selectivity through the use of copper-tris(triazole) ligand complexes. Org Lett 13:2754–2757
9. Pangborn AB, Giardello MA, Grubbs RH, Rosen RK, Timmers FJ (1996) Safe and convenient procedure for solvent purification. Organometallics 15:1518–1520

Chapter 4

Protecting-Group-Free Synthesis of Glycopolymers and Their Binding Assay with Lectin and Influenza Virus

Tomonari Tanaka, Tadanobu Takahashi, and Takashi Suzuki

Abstract

Typical synthetic methods for glycopolymers are laborious and require multistep processes, including protection and deprotection steps. Here we describe a facile protecting-group-free synthetic approach to glycopolymers bearing oligosaccharides from free saccharides by direct azidation and click chemistry methods, followed by reversible addition-fragmentation chain transfer polymerization. This method can be applied not only to mono- and disaccharides, but also to large biologically relevant oligosaccharides having sialic acids. Due to the glycocluster effect, the glycopolymers strongly bind with the corresponding lectin and influenza A virus, as analyzed by the quartz crystal microbalance method and hemagglutination inhibition assay.

Key words Glycopolymer, Cluster effect, Glycosyl azide, Click chemistry, RAFT polymerization, Lectin, Influenza virus

1 Introduction

Glycopolymers are synthetic polymers with pendant saccharides and have received much attention in many fields, e.g., polymer chemistry, materials sciences, and biomedicine. Saccharide-protein interactions are generally weak, but these interactions are amplified by multivalent forms of glycopolymers; this amplification is called the "glycocluster effect." However, these glycopolymers require multistep syntheses involving laborious steps, including the protection and deprotection of hydroxy groups on saccharides. Typically, the glycomonomers for the preparation of glycopolymers are synthesized by first protecting all the hydroxy groups of the starting free saccharides, activating their anomeric position, introducing a polymerizable group such as a vinyl group at the anomeric position, and finally deprotection. The resulting glycomonomers are used in the polymerization reaction, giving rise to glycopolymers.

We recently reported the facile protecting-group-free synthesis of glycopolymers from the free saccharides Lac, SA-Lac, and

Xue-Long Sun (ed.), *Macro-Glycoligands: Methods and Protocols*, Methods in Molecular Biology, vol. 1367, DOI 10.1007/978-1-4939-3130-9_4,

N-glycan [1]. Here we first describe the synthesis of glycomonomers having an acrylamide (AAm) group at the anomeric position by the direct azidation of free saccharides using 2-chloro-1,3-dimethylimidazolinium chloride (DMC) as a condensing agent, followed by copper(I) catalyzed azide-alkyne cycloaddition with *N*-propargyl acrylamide, which reaction is called "click chemistry." The resultant glycomonomers were copolymerized with AAm by reversible addition-fragmentation chain transfer (RAFT) polymerization. Next, the glycopolymers were immobilized on a gold-coated quartz crystal microbalance (QCM) sensor to analyze their binding behavior with the corresponding lectin. Finally, we describe the results of assaying binding of the glycopolymers with influenza A virus by the hemagglutination inhibition (HI) method.

2 Materials

2.1 Chemicals

1. Lactose (Lac).
2. 6′-Sialyllactose (SA-Lac).
3. *N*-glycan, biantennary complex-type sialyloligosaccharide, is produced from hen egg yolk glycopeptide hydrolyzed with endo-β-*N*-acetylglucosaminidase from *Mucor hiemalis* (Endo-M) as described in [2].
4. Pullulan (GPC STANDARD, Shodex).
5. *N*-propargyl acrylamide is synthesized from acryloyl chloride and propargyl amine in the presence of triethylamine as described in [3].
6. Tris[(1-benzyl-1*H*-1,2,3-triazol-4-yl)methyl]amine (TBTA) is synthesized as a stabilizing ligand for copper(I) from tripropargylamine, benzyl azide, and tetrakis(acetonitrile)copper(I) hexafluorophosphate as described in [4].
7. 2-(Benzylsulfanylthiocarbonylsulfanyl)ethanol as a chain transfer agent (CTA) is synthesized from 2-mercaptoethanol, carbon disulfide, and benzyl bromide as described in [5].
8. Acrylamide (AAm) is used after recrystallization from chloroform/methanol 10/3 (v/v).

2.2 Proteins and Viruses

1. Peanut agglutinin from *Arachis hypogaea* (PNA) (Wako Pure Chemical Industries, Japan).
2. *Sambucus sieboldiana* agglutinin (SSA) (Wako Pure Chemical Industries, Japan).
3. Bovine serum albumin (BSA).
4. The influenza A virus strain A/Memphis/1/1971 (H3N2) is propagated and purified as described in [6] (*see* **Note 1**).
5. Guinea-pig erythrocyte from fresh blood of guinea-pig.

2.3 Chromatography

2.3.1 TLC

1. TLC is performed on Silica Gel 60 Plate F254 (Merck).
2. Spots are visualized by spraying with a coloring reagent, then heating.

2.3.2 Ion-Exchange Column Chromatography

1. Ion-exchange resin column chromatography is performed in water on Amberlite IR-120B NA (Organo) previously activated with 1 M NaOH [7].

2.3.3 Gel Permeation Chromatography for Purification of N-Glycan Derivatives

1. Gel permeation chromatography (GPC) is performed on Sephadex G25 super fine (GE Healthcare) in 0.001 % NH_3 aq.

2.3.4 Silica Gel Column Chromatography for Purification of Saccharide Derivatives

1. Silica gel column chromatography is performed on silica gel 60 (70-230 mesh) (Nacalai Tesque) in aqueous CH_3CN.

2.3.5 Analytical GPC

1. Analytical GPC of the glycopolymers is carried out using an HPLC system; Shodex OHpak SB-804 HQ column (ϕ8.0 × 300 mm); mobile phase of 20 mM phosphate buffer (pH 7.0); flow rate of 0.5 mL/min; and RI detection at 30 °C.
2. Pullulan samples are used as standards.

3 Methods

Protecting-group-free synthesis of glycopolymers and their biological assay are described below. A typical synthetic procedure for glycopolymers from free saccharides is as follows (Fig. 1). β-Glycosyl azides **1** are directly synthesized from free saccharides using DMC, sodium azide, and a base agent in water, and reacted with *N*-propargyl acrylamide in the presence of a catalytic amount of copper(II) sulfate pentahydrate, L-ascorbic acid sodium salt, and TBTA in aqueous DMF. The resulting products **2**, which are AAm derivatives having triazole-linked oligosaccharide residues, are obtained and purified by silica gel column chromatography. The glycomonomers **2** are subjected to RAFT co-polymerization with AAm to obtain glycopolymers **3** in DMSO at 35 °C using 2,2-azobis(4-methoxy-2,4-dimethylvaleronitrile) (V-70) and 2-(benzylsulfanylthiocarbonylsulfanyl)ethanol as the initiator and CTA, respectively. The glycopolymers **3** bearing oligosaccharides are isolated after dialysis. The saccharide unit ratio in the product polymers is slightly lower than the glycomonomer ratio in the feed. This method can be applied not only to mono- and disaccharides, but also to large biologically relevant oligosaccharides having sialic acids.

The interaction between glycopolymers and proteins can be detected by the QCM method [8]. The thiol-terminated glycopolymers **4**, which are easily prepared by reducing the trithiocarbonate terminal group of **3** with sodium borohydride, are immobilized on

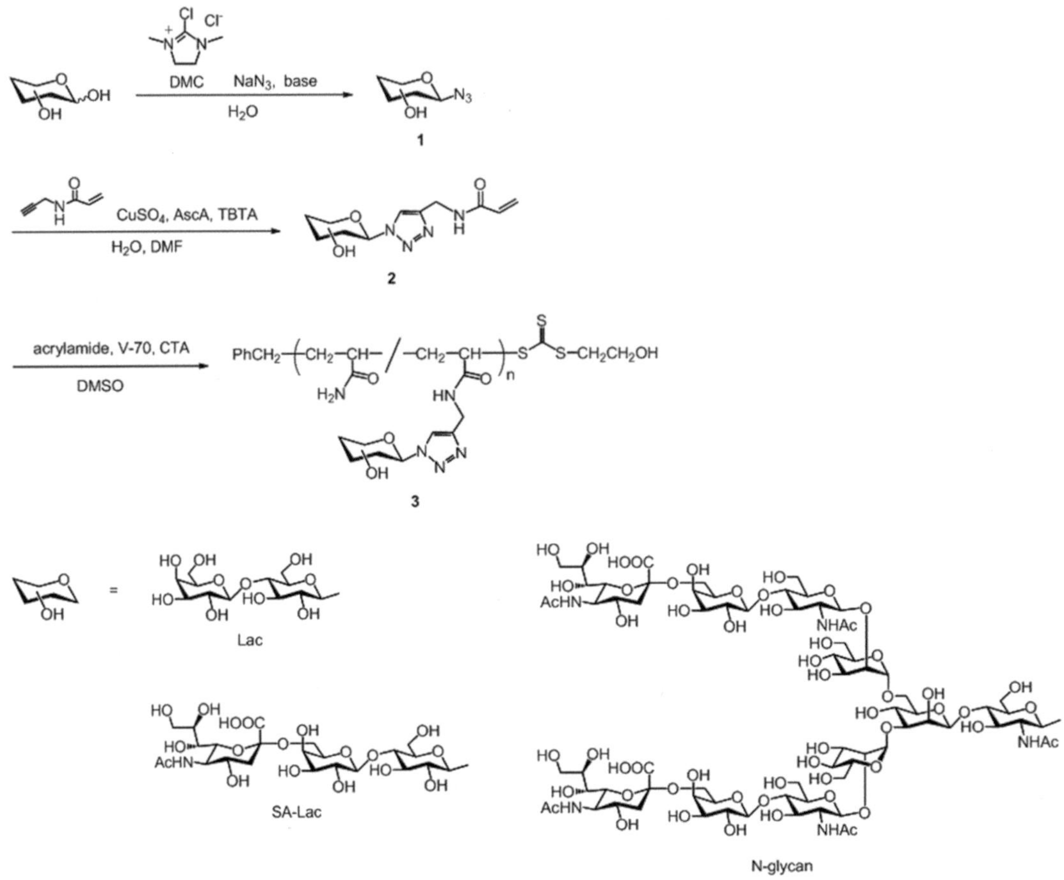

Fig. 1 Protecting-group-free synthesis of glycopolymers from free saccharides

a gold-coated QCM sensor via Au-S bond formation and subjected to a binding test with proteins in aqueous solution (Fig. 2). When the corresponding lectin is added to the QCM sensor immobilized with the glycopolymer, the frequency decreases, indicating an increase in mass due to the saccharide-bound lectin. The addition of PNA and SSA, which specifically recognize the β-galactoside and α2–6 sialylgalactoside residues, respectively, to a QCM sensor immobilized with the Lac, SA-Lac, and N-glycan-bearing glycopolymers significantly decreased the frequency, indicating strong saccharide-protein interactions (Fig. 3). When BSA was added to the QCM sensor immobilized with the glycopolymer, no decrease in frequency was observed. The binding constant, e.g., the association constant (K_a), can be estimated from the QCM analysis. It was previously reported that the K_a value for the binding of lectin and free saccharide is on the order of 10^3 M^{-1}.[9] The K_a values for the binding of glycopolymer and the corresponding lectin are on the order of 10^6–10^7 M^{-1}. These substantially higher K_a values for glycopolymer are attributed to the glycocluster effect where the lectin-saccharide

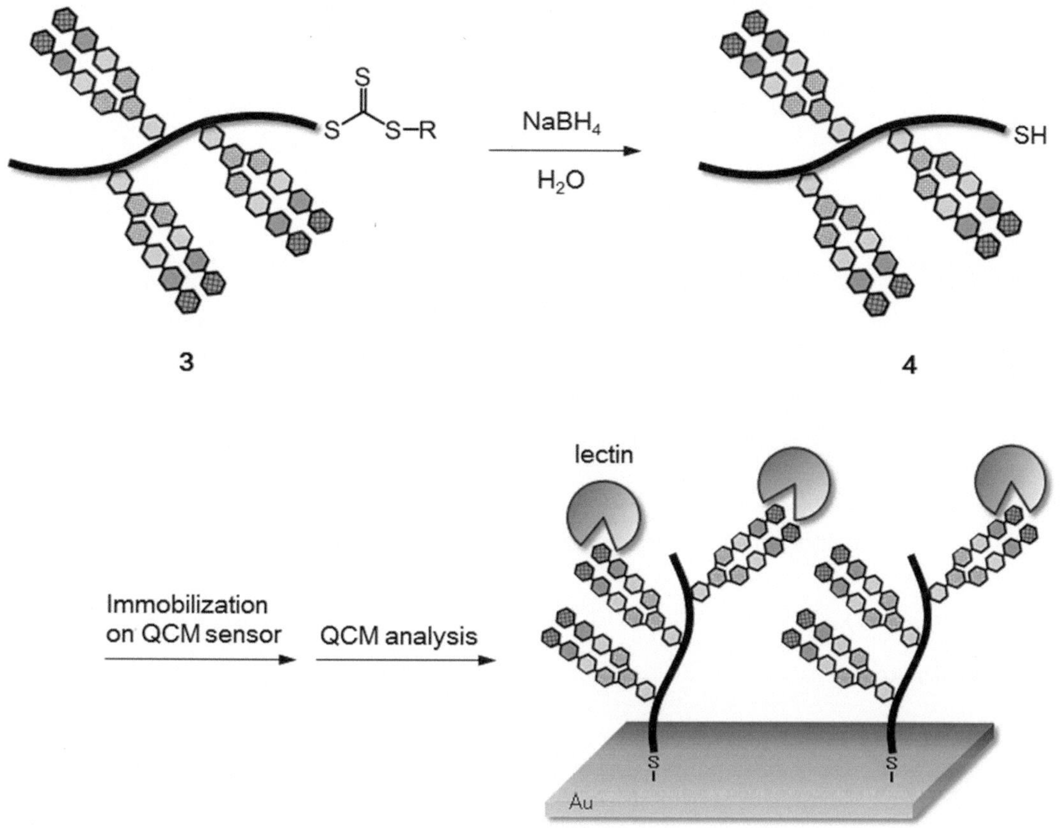

Fig. 2 Immobilization and assay of glycopolymers on a QCM sensor

interactions are amplified by the multivalency of the saccharides on the synthetic polymer.

The binding of glycopolymers having sialic acids with influenza A virus is demonstrated by the HI assay. Hemagglutination is observed with glycopolymers bearing sialyloligosaccharides (Fig. 4). The minimum concentration required to obtain a positive result using N-glycan-bearing polymer is much lower than that of SA-Lac-bearing polymer. No activity is observed with a glycopolymer that lacks a sialic acid residue, e.g., Lac-bearing glycopolymer. This assay is simple and straightforward and the results are confirmed visually.

3.1 Synthesis of Glycomonomers

3.1.1 Synthesis of Glycosyl Azides

1. Add an appropriate amount (*see* Table 1) of DMC, free saccharide, base agent, and sodium azide into water at 0 °C (*see* **Note 2**).
2. Stir the resulting mixture at 0 °C for 1 h.

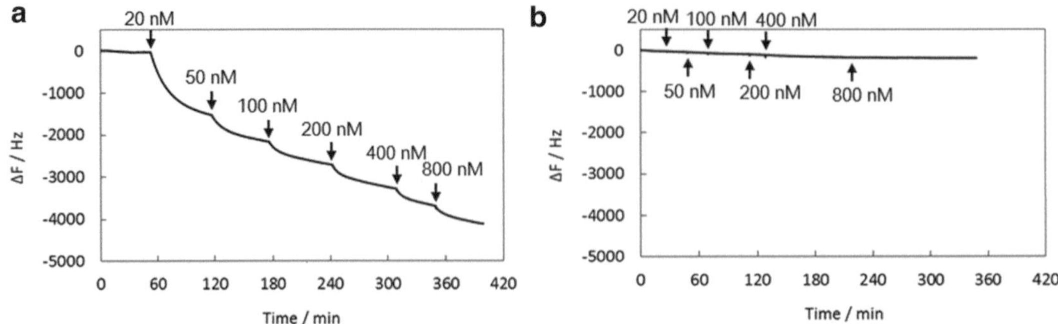

Fig. 3 QCM analysis of glycopolymer-protein interaction. (**a**) SA-Lac-PAAm/SSA, (**b**) SA-Lac-PAAm/BSA

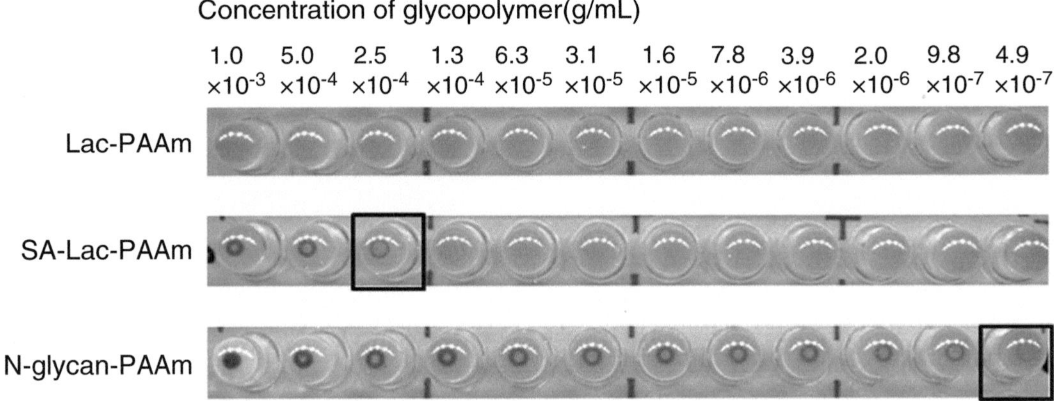

Fig. 4 HI assay of glycopolymers against human influenza virus A/Memphis/1/1971 (H3N2). The squares show the minimum concentration required for HI activity

Table 1
Synthesis of β-glycosyl azides 1 from free saccharide

Free saccharide (mM)	DMC (equiv.)	Base (equiv.)	NaN_3 (equiv.)	Time (h)	Temp. (°C)	Yield (%)
Lac (250)	3	DIPEA[a] (9)	10	1	0	quant.
SA-Lac (250)	3	DIPEA (9)	10	1	0	92
N-glycan (40)	20	2,6-lutidine (40)	62.5	24	r.t.	64

[a] *N*,*N*-diisopropylethylamine

3. Concentrate the reaction mixture in vacuo at 40 °C to afford a residue.
4. Add DMF (excess to dissolve the product) to the residue and filter the solution using filter paper to remove the solid.
5. Collect the filtrate and concentrate in vacuo at 40 °C to afford a residue.
6. Dissolve the residue in water (excess to dissolve the product) and extract with dichloromethane (same quantity of water). Water layer is used for purifying by ion-exchange column chromatography or GPC.
7. β-Lactosyl azide (Lac-N_3) and β-sialyllactosyl azide (SA-Lac-N_3) are purified by ion-exchange column chromatography performed in water on Amberlite IR-120B NA (Organo) previously activated with 1 M NaOH.
8. Disialo-complex-type oligosaccharyl azide (N-glycan-N_3) was purified by GPC using Sephadex G25 super fine (ϕ20 × 1100 mm) in 0.001 % NH_3 aq.
9. Add a metal scavenger (SiliaMets® Imidazole (SiliCycle), five equiv. for removing copper) to the aqueous solution of product and stir overnight at room temperature (*see* **Note 3**).

3.1.2 Synthesis of Glycomonomers by Click Chemistry

1. Add *N*-propargyl acrylamide, copper(II) sulfate pentahydrate, L-ascorbic acid sodium salt, and TBTA to a solution of β-glycosyl azide **1** in aqueous DMF (H_2O/DMF = 1/1) (*see* Table 2).
2. Stir the resulting mixture for 24 h at room temperature.
3. Concentrate the reaction mixture in vacuo at 40 °C to afford a residue.
4. Purify the product by silica gel column chromatography (CH_3CN/H_2O = 5/1; Lac and SA-Lac, 5/2; N-glycan).
5. Remove the metal scavenger by filtration using filter paper.
6. Collect the filtrate and concentrate in vacuo at 40 °C and freeze-dry to give the glycomonomer **2**.

Table 2
Synthesis of glycomonomers 2 by click chemistry

Glycosyl azide1[a]	Acetylene/$CuSO_4$/AscA[b]/TBTA (equiv.)	Yield (%)
Lac-N_3	1.0/0.1/0.2/0.1	94
SA-Lac-N_3	1.1/0.1/0.2/0.1	66
N-glycan-N_3	1.2/0.1/0.2/0.1	70

[a]50 mM
[b]L-ascorbic acid sodium salt

Table 3
Synthesis of glycopolymers 3 by RAFT polymerization[a]

Glycomonomer 2	Feed molar ratio of 2/AAm	Conv. (%)[b]	Glycopo er3	Yield (%)[c]	Mn[b]	Mw/Mn[d]	Saccharide ratio in polymer (%)[b]
Lac-AAm	1/9	71	Lac-PAAm	65	23,000	1.16	7.2
SA-Lac-AAm	1/9	72	SA-Lac-PAAm	72	23,000	1.81	7.7
N-glycan-AAm	1/9	87	N-glycan-PAAm	87	33,000	2.66	7.3

[a]At a molar ratio of total monomer/CTA/V-70 = 750/5/1 in DMSO at 35 °C for 24 h
[b]Determined by ^{1}H NMR
[c]Isolated yield after dialysis
[d]Determined by GPC

3.2 Synthesis of Glycopolymers

1. Dissolve the glycomonomer **2**, AAm, V-70, and CTA in DMSO (0.25 mL) and degas it by three freeze-thaw cycles in a glass tube (*see* Table 3).
2. Seal the glass tube under vacuum and heat it at 35 °C for 24 h.
3. Purify the product by dialysis (Spectra/Por 7 MWCO 3500 (Spectrum)) against water and freeze-dry to give the corresponding glycopolymer **3**.

3.3 Biological Assay of Glycopolymers

3.3.1 Detection of Glycopolymer-Lectin Interaction by QCM

1. To reduce the trithiocarbonate terminal group of the glycopolymer, dissolve the glycopolymers **3** (9 mg) and sodium borohydride (2 mg) individually in water (100 μL) and stir for 1 h at room temperature.
2. Purify the corresponding thiol-terminated glycopolymer **4** by dialysis (Spectra/Por 7 MWCO 3500) against water and followed by freeze-dry.
3. Drop a PBS solution of thiol-terminated glycopolymer **4** (0.01 mg/mL) on the gold-coated QCM sensor to immobilize the glycopolymer via Au-S bonds.
4. After 90 min, wash the QCM sensor with water and set it on a QCM system.
5. Add a lectin (20–800 nM) in PBS to the QCM sensor. Record the frequency decreased when the corresponding lectin added (*see* Fig. 3).

3.3.2 HI Assay

1. Add 50 μL of 0.5 % (v/v) guinea-pig erythrocyte suspension in PBS to each well (25 μL) of a U-bottom 96-well microtiter

plate containing a twofold serial dilution of influenza A virus in PBS.

2. After incubation at 4 °C for 2 h, the maximum dilution of the virus showing hemagglutination is defined as the number of viral hemagglutination units.
3. Add 25 μL of influenza A virus suspension (four hemagglutination units) in PBS to each well (25 μL) containing the glycopolymers **3** in a twofold serial dilution in PBS using a U-bottom 96-well microtiter plate.
4. After incubation at 4 °C for 1 h, add 50 μL of 0.5 % (v/v) guinea-pig erythrocyte suspension to the plates and stored for 2 h at 4 °C.
5. The maximum dilution of the glycopolymer showing complete inhibition of hemagglutination is defined as the HI titer (highlighted by square borders; *see* Fig. 4).

4 Notes

1. Influenza A virus was propagated in the allantoic sacs of 10-day-old embryonated hen's eggs for 2 days at 34 °C and purified by sucrose density gradient centrifugation.
2. Sodium azide may explosively decompose on exposure to heat, shock, or friction. Do not use a metal spatula.
3. After silica gel column chromatography, the remaining copper can be removed by using a metal scavenger.

Acknowledgment

This work was financially supported by JSPS KAKENHI Grant No. 25810075.

References

1. Tanaka T, Ishitani H, Miura Y, Oishi K, Takahashi T, Suzuki T, Shoda S, Kimura Y (2014) Protecting-group-free synthesis of glycopolymers bearing sialyloligosaccharide and their high binding with the influenza virus. ACS Macro Lett 3:1074–1078
2. Umekawa M, Higashiyama T, Koga Y, Tanaka T, Noguchi M, Kobayashi A, Shoda S, Huang W, Wang L-X, Ashida H, Yamamoto K (2010) Efficient transfer of sialo-oligosaccharide onto proteins by combined use of a glycosynthase-like mutant of Mucor hiemalis endoglycosidase and synthetic sialo-complex-type sugar oxazoline. Biochim Biophys Acta Gene Sub 1800:1203–1209
3. Wipf P, Aoyama Y, Benedum TE (2004) A practical method for oxazole synthesis by cycloisomerization of propargyl amides. Org Lett 6:3593–3595
4. Chan TR, Hilgraf R, Sharpless KB, Fokin VV (2004) Polytriazoles as copper(I)-stabilizing ligands in catalysis. Org Lett 6:2853–2855
5. Hales M, Barner-Kowollik C, Davis TP, Stenzel MH (2004) Shell-cross-linked vesicles synthe-

sized from block copolymers of poly(D, L-lactide) and poly(N-isopropyl acrylamide) as thermoresponsive nanocontainers. Langmuir 20:10809–10817

6. Suzuki Y, Nagao Y, Kato H, Matsumoto M, Nerome K, Nakajima K, Nobusawa E (1986) Human influenza A virus hemagglutinin distinguishes sialyloligosaccharides in membrane-associated gangliosides as its receptor which mediates the adsorption and fusion processes of virus infection. J Biol Chem 261:17057–17061
7. Vinson N, Gou Y, Becer CR, Haddleton DM, Gibson MI (2011) Optimised 'click' synthesis of glycopolymers with mono/di- and trisaccharides. Polym Chem 2:107–113
8. Ebara Y, Okahata Y (1994) A kinetic study of concanavalin A binding to glycolipid monolayers by using a quartz-crystal microbalance. J Am Chem Soc 116:11209–11212
9. Miura Y, Ikeda T, Kobayashi K (2003) Chemoenzymatically synthesized glycoconjugate polymers. Biomacromolecules 4:410–415

Chapter 5

Carbohydrate-Based Initiators for the Cationic Ring-Opening Polymerization of 2-Ethyl-2-Oxazoline

Christine Weber, Michael Gottschaldt, Richard Hoogenboom, and Ulrich S. Schubert

Abstract

The advancement in the field of living and controlled polymerization techniques provides the opportunity for careful bottom-up design of polymers for biomedical applications according to their specific needs. This contribution describes the detailed methodology to functionalize poly(2-ethyl-2-oxazoline), a polymer with properties very similar to polyethylene glycol, in a stereo-selective manner with a range of carbohydrates that can serve as biological targeting units. The obtained building blocks can subsequently be applied for the synthesis of more complex polymeric architectures.

Key words Cationic polymerization, Ring-opening polymerization, Telechelic polymers, End-functional polymers, Carbohydrate, Glycopolymers, Poly(2-oxazoline)

1 Introduction

Poly(2-oxazoline)s (POx) are a class of polymers that can be obtained by living cationic ring-opening polymerization (CROP). The fact that the two water-soluble POx, i.e., poly(2-methyl-2-oxazoline) and poly(2-ethyl-2-oxazoline) (PEtOx), reveal similar properties as the widely used poly(ethylene glycol) has driven forward the research directed toward their application in the biomedical field. In view of this, it is advantageous that the livingness of the CROP provides the possibility to (orthogonally) functionalize both end-groups of the POx chain with molecules that have biological functions, such as carbohydrates, or with other functional moieties (Fig. 1). The latter can be employed for the embedding of the "stealth polymer/cell target building blocks" into larger polymeric architectures, with focus on the development of designed macromolecules for drug delivery.

We report the development of suitable triflate initiators, which are based on protected glucose, galactose and fructose, for the

Xue-Long Sun (ed.), *Macro-Glycoligands: Methods and Protocols*, Methods in Molecular Biology, vol. 1367,
DOI 10.1007/978-1-4939-3130-9_5,

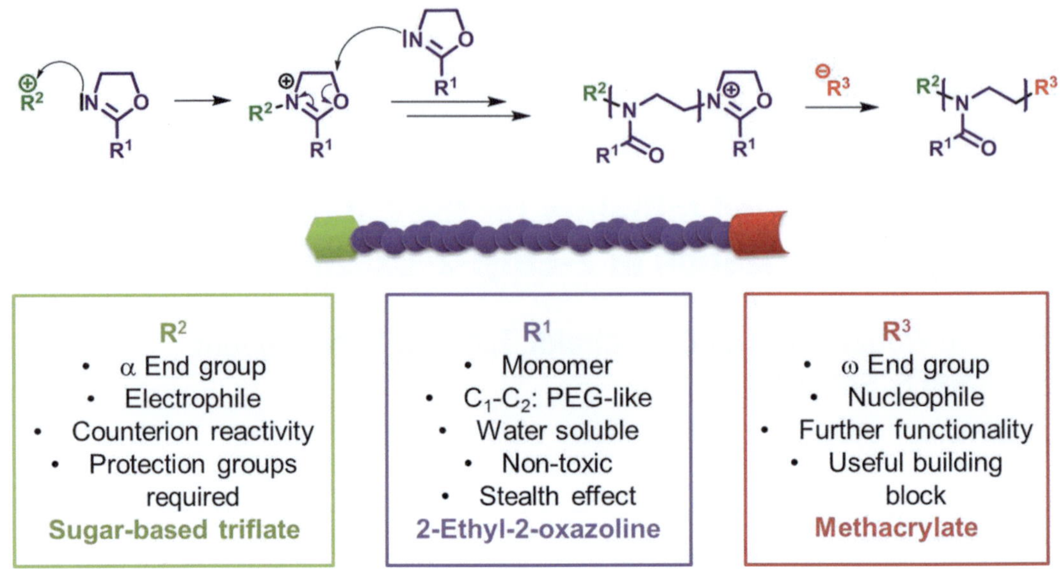

Fig. 1 Schematic representation of the mechanism of the cationic ring-opening polymerization (CROP) of 2-oxazolines and requirements for initiators and end-functionalization agents with focus on the systems discussed in this contribution

CROP of EtOx [1]. To exemplify how these α-end-functional POx can be further functionalized on the ω-chain end, we discuss the synthesis of a methacrylate end-functionalized macromonomer. This approach enables their further polymerization by very robust controlled radical polymerization techniques, such as the reversible addition-fragmentation chain transfer (RAFT) polymerization, making it possible to exploit the wide range of commercially available (meth)acrylate monomers in combination with the POx-based building blocks [2, 3].

2 Materials

2.1 Chemicals

1. 1,2:3,4-Di-*O*-isopropylidene-6-*O*-trifluormethanesulfonyl-α-D-galactopyranose (DIPGalTf) is synthesized from 1,2:3,4-di-*O*-isopropylidene-α-D-galactopyranose (Aldrich) and trifluoromethanesulfonic acid anhydride (Fluka) as described in literature [4] (*see* **Notes 1** and **2**).
2. 2,3:4,5-Di-*O*-isopropylidene-1-*O*-trifluoromethansulfonyl-D-fructopyranose (DIPFruTf) is synthesized from 2,3:4,5-di-*O*-isopropylidene-β-D-fructopyranose (Carbosynth) and trifluoromethanesulfonic acid anhydride (Fluka) as described in literature [5] (*see* **Notes 1** and **2**).

3. 2-Trifluoromethylsulfonyl oxyethyl 2,3,4,6-tetra-*O*-acetyl-β-D-glucopyranoside (Ac_4GlcTf) is synthesized in a two-step synthesis from 2,3,4,6-tetra-*O*-acetyl-acetobromo-α-D-glucose (97 %, AlfaAesar), ethylene glycol, and trifluoromethanesulfonic acid anhydride (Fluka) as described in literature [1] (*see* **Notes 1** and **3**).
4. 2-Ethyl-2-oxazoline (EtOx) (Acros) is distilled over BaO and kept under argon.
5. Acetonitrile (Acros, extra dry) is stored under argon.
6. Methacrylic acid (MAA) (99 %, Aldrich) is used as received. The use of inhibitor remover is not necessary because the inhibitor will be removed during the purification process of the macromonomer in later stages.
7. Triethylamine (NEt_3) is kept over potassium hydroxide, distilled, and stored under argon.
8. 2,2′-Azobis(2-methylpropionitrile) (AIBN) (98 %, Fluka) is recrystallized once from methanol.
9. 2-Cyanopropyl dithiobenzoate (CPDB) (97 %, Aldrich).
10. Ethanol (Sigma-Aldrich, absolute).
11. 0.5 M Sodium methoxide solution in methanol (Sigma).
12. Methanol (extra dry, Acros).

2.2 Other Commercial Reagents and Solvents

1. BioBeads SX-1 (BioRad).
2. Tetrahydrofuran (THF, redistilled).

2.3 Chromatography

2.3.1 Thin-Layer Chromatography

1. Thin-layer chromatography (TLC) is performed on silica gel-covered plates in ethyl acetate/hexane in a volume ratio 1/1.
2. Spots are visualized by dipping the plate in a mixture of methanol/sulfuric acid (ratio 10/1) and subsequent burning with hot air.

2.3.2 Preparative Size-Exclusion Chromatography

1. Preparative size-exclusion chromatography (SEC) is performed on a column filled with BioBeads SX-1 using THF as eluent.
2. Prior to combination of the desired fractions, the absence of unreacted macromonomer is confirmed by analytical SEC (*see* **Note 4**).

2.3.3 Analytical Size-Exclusion Chromatography

1. Analytical SEC for the linear polymers is performed using a PSS-SDV-linear M 5 μm column at 40 °C with a mobile phase composed of chloroform, triethylamine, and isopropanol (94:4:2) at a flow rate of 1 mL/min (RI detection).
2. Analytical SEC for the comb polymers is performed on a PSS Gram30 and a PSS Gram1000 column in series, whereby *N*,*N*-

dimethylacetamide (DMAc) with 2.1 g/L of LiCl is applied as an eluent at 1 mL/min flow rate and the column oven is set to 40 °C (RI detection).

3. The systems are calibrated with polystyrene (2000–88,000 g/mol) standards.
4. Data analysis is performed using the WinGPC software from PSS (Mainz, Germany).

2.4 Nuclear Magnetic Resonance Spectroscopy

1. ^{1}H and ^{13}C nuclear magnetic resonance (NMR) spectra are recorded in $CDCl_3$ or DMSO-*d6* on a Bruker Avance 300 MHz using the residual solvent resonance as an internal standard (*see* **Note 5**).
2. ^{19}F NMR spectra are measured on a Bruker Avance 200 MHz spectrometer.

2.5 Mass Spectrometry (MS)

2.5.1 Matrix-Assisted Laser Desorption Ionization Time-of-Flight

1. Matrix-assisted laser desorption ionization time-of-flight (MALDI TOF MS) is measured on a Ultraflex III TOF/TOF (Bruker Daltonics, Bremen, Germany).
2. Samples, the matrix *trans*-2-[3-(4-*tert*-butylphenyl)-2-methyl-2-propenylidene]malononitrile and the ionization salt sodium iodide are spotted on the target using the dried droplet method [6].
3. The instrument is calibrated with an external PMMA standard from PSS Polymer Standards Services GmbH (Mainz, Germany).

2.5.2 Electro-Spray Ionization Time-of-Flight MS

1. Electro-spray ionization time-of-flight (ESI TOF) MS measurements are performed with a micrOTOF Q-II (Bruker Daltonics) mass spectrometer.
2. The ESI TOF MS instrument is calibrated in the *m/z* range from 50 to 3000 using a calibration standard (Tunemix solution) supplied from Agilent.

3 Methods

When designing initiators for the CROP of EtOx, one must consider that the cationic polymerization mechanism prohibits the use of nucleophilic functional moieties that might cause chain transfer or can quench the oxazolinium species. Consequently, triflate functional isopropylidene or acetyl protected carbohydrates based on glucose, galactose and fructose were applied (Fig. 2). The triflate moiety acts as leaving group upon the attack of the lone pair of the nitrogen atom in the EtOx monomer ring during the initiation step of the CROP. Subsequently, it represents the counter ion during the polymerization process. Although the corresponding sugar tosylates are more easily prepared and much more

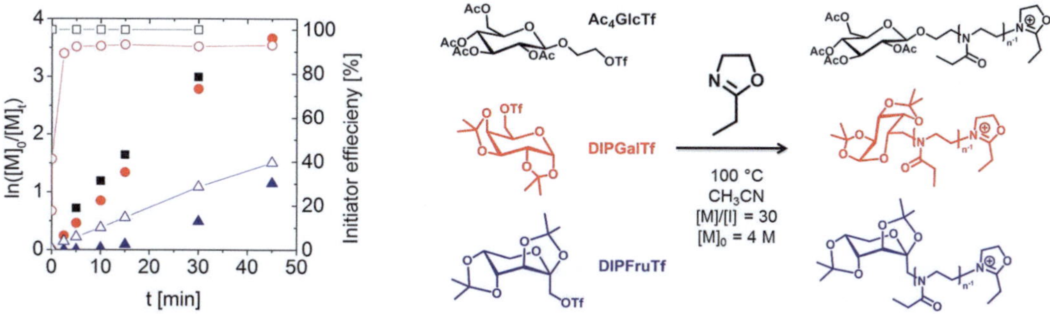

Fig. 2 *Right*: Schematic representation of the carbohydrate initiators employed for the CROP of EtOx and structure of the resulting α-end-functional polymers. *Left*: Semilogarithmic kinetic plot obtained from the corresponding kinetic studies using these initiators. Ac_4GlcTf (*black squares*, *kp* = 0.0121 L/(mol*s)), DIPGalTf (*red circles*, *kp* = 0.0124 L/(mol*s)), DIPFruTf (*blue triangles*). The *open symbols* indicate the initiation efficiencies determined by ^{19}F NMR spectroscopy

stable than the triflates discussed here, it is exactly their stability that causes slow initiation of the CROP and, thus, prevents the control over the molar mass of the resulting PEtOx [1].

As evident from the linear semilogarithmic kinetic plot in Fig. 2, both Ac_4GlcTf and DIPGalTf act as fast initiators for the CROP of EtOx, resulting in narrow molar mass distributions of the formed PEtOx. The polymerization rate coefficient *kp* can be determined from the slope of the linear fit using the equation specified in **Note 9**. Due to the fact that the polymerization conditions were the same for both initiators and the counter ion was triflate in both cases, the respective *kp* values are very similar (*see* Fig. 2 and **Note 6**). The attachment of both protected sugars to the polymer chains was confirmed by NMR spectroscopy as well as ESI and MALDI TOF MS, taking advantage of the distinct peaks resulting from the sugar bearing fragments in tandem MS investigations [7]. Although the DIPFruTf was capable of initiating the CROP as well and its attachment to the PEtOx could be confirmed in the same manner, a slow initiation was observed, presumably due to steric hindrance, leading to the typical, rather broad, molar mass distributions with low molar mass tailing.

Subsequently, the sugar moieties at the α-chain ends of the PEtOx can be deprotected applying the standard conditions for isopropylidene or acetyl groups, respectively [1]. Alternatively, the living oxazolinium ω-chain end of the PEtOx can be utilized to introduce another functional moiety by direct quenching of the CROP with suitable nucleophiles, such as methacrylate anions. These can be formed in situ by simple addition of MAA and NEt_3 as base to the solution after polymerization (Fig. 3). The resulting PEtOx with both a protected glucose and a methacrylate end functionality, represents a macromonomer suitable for RAFT polymerization. Thereby, a comb polymer with a polymethacrylate

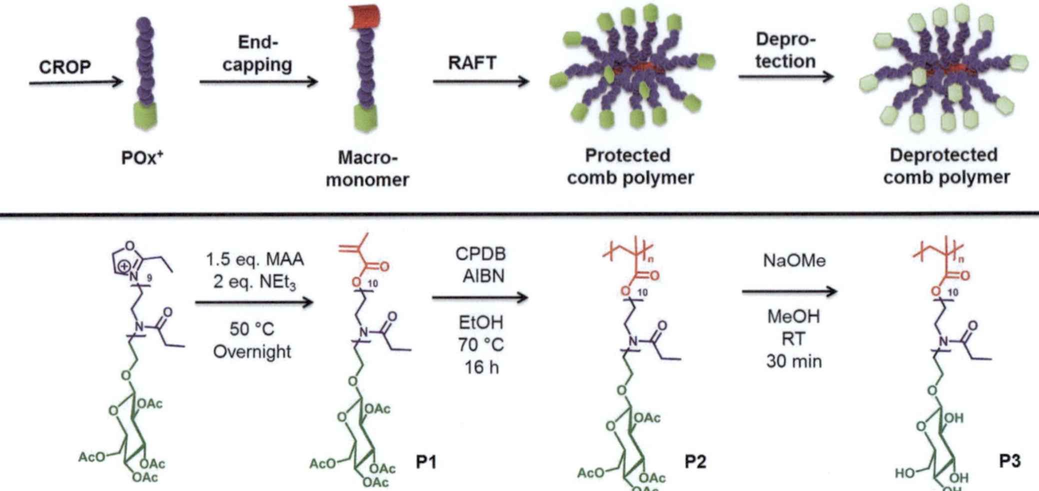

Fig. 3 Schematic representation of the synthesis route towards a comb polymer with polymethacrylate backbone and PEtOx side chains that is selectively functionalized with glucose at the side chain ends

backbone and PEtOx side chains, which carry the protected glucose moieties at the chain ends, is formed. The subsequent careful deprotection yielded the corresponding "sugar brush". Due to the tight comb architecture, where each repeating unit of the backbone carries a PEtOx chain, the entanglement of the polymer chains is restricted and the sugar moieties are exposed to the outside of the macromolecular architecture.

To proof the successful synthesis, careful characterization has to be performed throughout the entire synthetic route. Due to the rather low molar mass of the macromonomer **P1** (a decamer) extensive MS investigations could be performed [1, 7], but a combination of SEC and NMR measurements is more suited as soon as the comb polymer must be analyzed (Fig. 4). The SEC elugrams clearly show the absence of unconverted macromonomer in the protected comb polymer **P2** after purification by preparative SEC. The shift of the narrow and unimodal SEC trace towards lower elution volume indicates a successful RAFT polymerization, while the ^{1}H NMR spectrum reveals that the PEtOx and the protected glucose moieties remained unaffected by the radical polymerization of the methacrylate moiety. The successful deprotection is confirmed by the disappearance of the corresponding acetyl signals in the ^{1}H NMR spectrum of **P3**, and the sugar proton signals can be identified when a HSQC NMR measurement is performed. However, only the combination with the unimodal SEC trace allows the conclusion that the comb structure itself remained unaffected by the deprotection of the sugar residues (*see* **Note 7**).

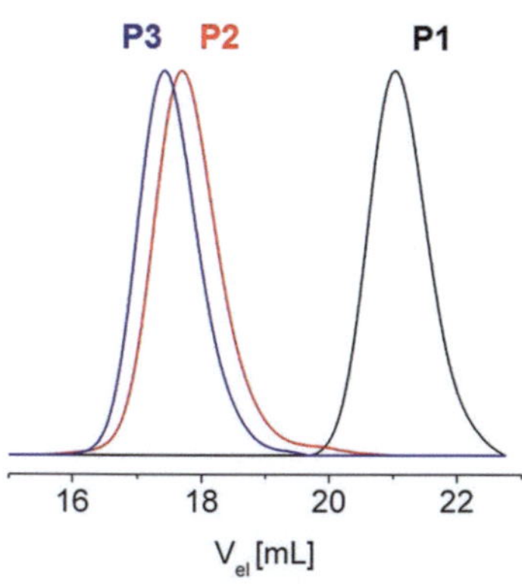

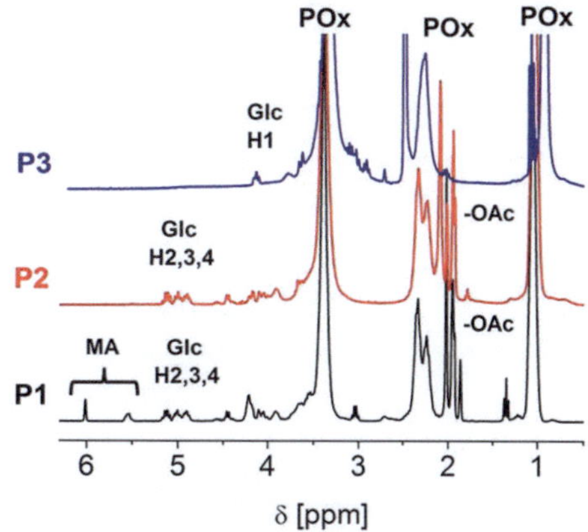

Fig. 4 *Left*: Normalized SEC elugrams (DMAc, RI detection) of the Ac_4GlcPEtOx-methacrylate macomonomer **P1**, the protected comb polymer **P2**, and the deprotected comb polymer **P3**. *Right*: ^{1}H NMR spectra of **P1–P3** (300 MHz, solvent: $CDCl_3$ for **P1** and **P2**, DMSO-d_6 for **P3**). Integration of the signals for "H2,3,4" vs. a "POx" signal is used to calculate the DP of the macromonomer **P1**. Integration of the signals for "H2,3,4" vs. a "MA" signal is used to calculate the degree of functionalization with the ω-methacrylate end group of the macromonomer **P1**. Reprinted (adapted) with permission from (C. Weber et al. *Macromolecules*, 2012, *45*, 46–55.). Copyright (2012) American Chemical Society

3.1 Cationic Ring-Opening Polymerization

1. All cationic ring opening polymerizations (CROP) are performed in an Initiator 60 microwave synthesizer from Biotage using the appropriate glass vials and caps with septa.
2. All glassware is heated to 110 °C overnight and allowed to cool under argon.
3. All solutions have to be prepared under inert atmosphere (*see* **Note 8**).

3.1.1 Kinetic Studies

1. Add 349 mg (0.667 mmol) of Ac_4GlcTf, 1983 mg (20 mmol) of EtOx and 3 mL of acetonitrile in a 20 mL microwave vial equipped with a magnetic stirrer (*see* **Note 9**).
2. Cap the vial and stir the stock solution shortly.
3. Transfer six aliquots of 700 μL of the stock solution via syringe through the septa of separate microwave vials (equipped with magnetic stirrers, suitable for reaction on 0.5–2 mL scale).
4. Place the vials into the autosampler of the microwave synthesizer.
5. The microwave synthesizer is programmed as follows: Temperature 100 °C, absorption level high. The reaction time differed for each vial, according to the times given in Fig. 2. Subsequent to reaction, each vial is cooled down automatically by a flow of nitrogen.

6. Open the vials and take samples for SEC (*see* Subheading 2.3.3) and NMR ($CDCl_3$) analysis.
7. The monomer conversions are calculated from the ^1H NMR spectra using the integrals of the peaks originating from the ring protons of EtOx (4.2 ppm and 3.8 ppm, respectively) and the PEtOx backbone signal at 3.4 ppm.
8. The initiation efficiency is calculated from the ^{19}F NMR signals of unreacted initiator (−75 ppm) and free triflate counter ions (−79 ppm).
9. For the galactose- and fructose-based initiators, 262 mg (0.667 mmol) of DIPGalTf or DIPFruTf, respectively, are used (*see* **Note 9**). All other steps are performed as described above.

3.1.2 Macromonomer (P1) Synthesis

1. Fill 524 mg (1 mmol) of Ac_4GlcTf, 991 mg (10 mmol) of EtOx and 1.5 mL of acetonitrile in a suitable microwave vial (reaction volume 2–5 mL) equipped with a magnetic stirrer.
2. The vial is reacted in the microwave synthesizer as follows: 1 min pre-stirring, temperature 100 °C, absorption level high, time 14 min.
3. Add 150 μL of MAA via syringe through the septum of the vial.
4. Add 300 μL of NEt_3 via syringe through the septum of the vial.
5. Place the vial in an oil bath set to 50 °C overnight.
6. Dissolve the reaction mixture in chloroform (100 mL) and wash it with saturated aqueous sodium bicarbonate solution (50 mL) and brine (50 mL).
7. Dry the organic phase over sodium sulfate.
8. Remove chloroform using an rotary evaporator (*see* **Notes 1** and **10**).
9. SEC: M_n = 1450 g/mol; Đ = 1.12. ^1H NMR: *see* Fig. 4, DP = 10; DF with methacrylate end group = 75 %.

3.2 Reversible Addition-Fragmentation Chain Transfer (RAFT) Polymerization

1. Dissolve 1 g (0.69 mmol) of **P1**, 3.8 mg (17 μmol) of CPDB and 0.7 mg of AIBN (4 μmol) in 1.4 mL of ethanol in a microwave vial (*see* **Note 11**).
2. Cap the vial and flush the solution with a gentle flow of argon for 30 min.
3. Place the vial in an oil bath, which is pre-heated to 70 °C, for 16 h.
4. The conversion of **P1** is determined from the peak integrals of macromonomer and comb polymer in the SEC elugram taken from the reaction solution (*see* **Note 12**).
5. Remove the volatiles under reduced pressure to afford the crude product. Dissolve the crude product in THF.
6. Remove residual macromonomer via preparative SEC.

7. Combine the desired fractions and concentrate the solution under reduced pressure to afford a viscous solution.
8. Drop the viscous solution to cold diethyl ether (50 mL) to precipitate the comb polymer **P2**.
9. Centrifuge the mixture, remove the supernatant and dry the residue under reduced pressure.
10. SEC: $M_n = 19{,}300$ g/mol; Đ = 1.14. ^{1}H NMR: *see* Fig. 4.

3.3 Deprotection

1. Dissolve 150 mg of **P2** in 1 mL of dry methanol in a microwave vial equipped with a magnetic stirrer.
2. Cap the vial and add 22 μL of 0.5 M sodium methoxide solution in methanol via syringe through the septum of the vial (*see* **Note 13**).
3. After 30 min, take a SEC sample to confirm the successful deprotection by the shift of the signal (*see* **Note 7**).
4. Neutralize the solution with diluted hydrochloric acid to pH 7.
5. Remove the volatiles under reduced pressure to afford the residue.
6. Dissolve the residue in 0.5 mL of ethanol. Remove the insolubles (sodium chloride) by filtering off using a syringe filter.
7. Drop the ethanolic solution into cold diethyl ether (20 mL) to precipitate comb polymer **P3**.
8. Centrifuge the mixture, remove the supernatant and dry the residue under reduced pressure.
9. SEC: $Mn = 23{,}600$ g/mol; Đ = 1.14. ^{1}H NMR: *see* Fig. 4. ^{1}H ^{13}C HSQC NMR: *see* Fig. 5.

4 Notes

1. Should be stored in a freezer at −18 °C.
2. Best results in the subsequent polymerizations are obtained when the initiator is used directly after synthesis. If stored for longer than 2 weeks, it is advised to confirm the purity of the initiator before usage by thin layer chromatography and/or to run a test polymerization on small scale.
3. Best results in the subsequent polymerizations are obtained when the initiator is used directly after synthesis. It should not be stored for longer than 1 week.
4. It may be possible to preselect the fractions that contain the comb polymer by the slight pink color originating from the dithiobenzoate end group.
5. For the end group confirmation using ^{13}C NMR, it is better to perform several measurements on the same sample and subse-

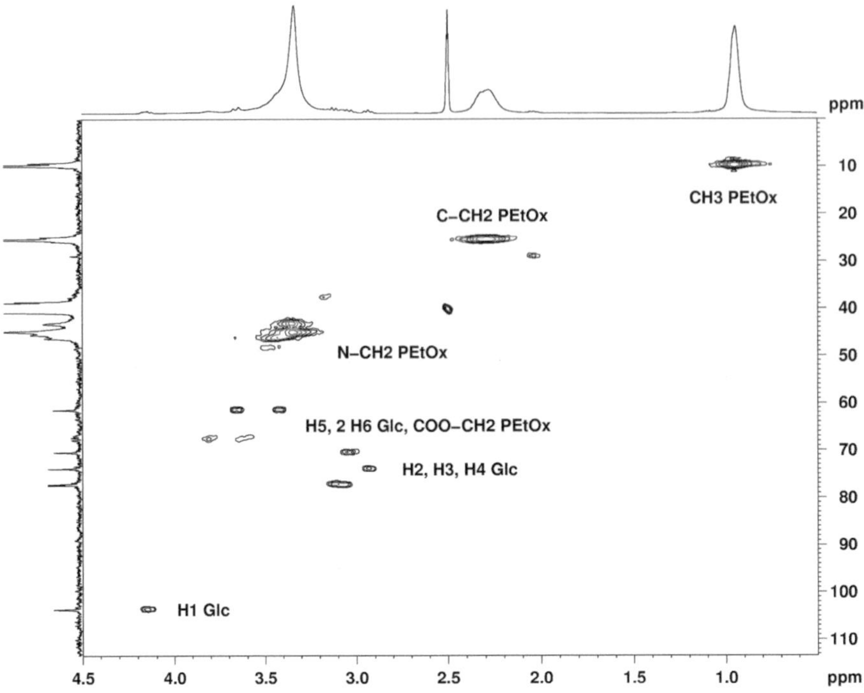

Fig. 5 ^{1}H ^{13}C HSQC NMR spectrum (DMSO-d_6, 400 MHz) of the PEtOx based comb polymer carrying deprotected glucose **P3**. Reprinted with permission from (C. Weber et al. *Macromolecules*, **2012**, *45*, 46–55.). Copyright (2012) American Chemical Society

quently use the option to show the sum of all spectra that is provided in the Bruker TopSpin software.

6. Even though the *kp* values are directly related to the utilized triflate counterions, it is better to determine them with the exact utilized microwave synthesizer equipment that is used. The values may slightly differ, depending on the type of microwave and the temperature calibration and readout that is used, or on the settings that are applied.
7. If the molar mass distribution becomes bimodal, some ester moieties that connect side chains and backbone were cleaved. If the SEC trace is not shifted, another 15 μL of sodium methoxide solution should be added.
8. It is possible to prepare the solutions while the microwave vials are constantly flushed with a flow of argon. As an alternative, the solutions can be prepared in a glove box. Depending on the quality of the inert gas, it may be necessary to install a column filled with common drying agent, such as phosphorous(V) oxide, calcium chloride, or potassium hydroxide.
9. This corresponds to a ratio of [monomer] to [initiator] $[M]_0/[I]_0$ of 30 and an initial monomer concentration $[M]_0$ of 4 mol/L. Both values may be varied but in this case the polym-

erization time has to be adjusted according to $\ln\left(\frac{[M]_0}{[M]_t}\right) = k_p \cdot [I]_0 \cdot t$.

10. The temperature should not exceed 40 °C and the evaporation of the chloroform should be done as fast as possible.
11. This corresponds to a [M]/[CPDB]/[AIBN] of 40/1/0.25 and a [M] of 0.5 mol/L. For the exact dosing of CPDB and AIBN, stock solutions were prepared and the appropriate volumes were added to reach the given amounts.
12. It should be confirmed that this method is applicable on the utilized SEC system by determination of the response factors of **P1** and **P2** and/or comparison with conversion data from ^{1}H NMR spectroscopy. If the DF of the macromonomer is not quantitative, this has to be taken into account.
13. It is important to use as little sodium methoxide as possible to prevent a cleavage of the side chains from the polymer backbone. As an rule of thumb, the catalytic amount $n(\text{NaOMe}) = 0.1$ $n(\text{Ac}_4\text{Glc})$ has proved to be very useful.

Acknowledgments

CW acknowledges the Carl-Zeiss foundation. The authors thank the Thuringian Ministry of Economic Affairs, Science and Digital Society (grants no. B515-07008 and B715-08011) and the Ernst-Abbe Stiftung for financial support of this study. We acknowledge Bruker Daltonics for their help and support.

References

1. Weber C, Czaplewska JA, Baumgaertel A, Altuntas E, Gottschaldt M, Hoogenboom R, Schubert US (2012) A sugar decorated macromolecular bottle brush by carbohydrate-initiated cationic ring-opening polymerization. Macromolecules 45(1):46–55
2. Weber C, Becer CR, Guenther W, Hoogenboom R, Schubert US (2010) Dual responsive methacrylic acid and oligo(2-ethyl-2-oxazoline) containing graft copolymers. Macromolecules 43(1):160–167
3. Weber C, Hoogenboom R, Schubert US (2012) Temperature responsive bio-compatible polymers based on poly(ethylene oxide) and poly(2-oxazoline)s. Prog Polym Sci 37(5):686–714
4. Streicher H, Busse H (2006) Building a successful structural motif into sialylmimetics: cyclohexenephosphonate monoesters as pseudo-sialosides with promising inhibitory properties. Bioorgan Med Chem 14(4):1047–1057
5. Card PJ, Hitz WD (1984) Synthesis of 1′-deoxy-1′-fluorosucrose via sucrose synthetase mediated coupling of 1-deoxy-1-fluorofructose with uridine-diphosphate glucose. J Am Chem Soc 106(18):5348–5350
6. Meier MAR, Adams N, Schubert US (2007) Statistical approach to understand MALDI-TOF-MS matrices: discovery and evaluation of new MALDI matrices. Anal Chem 79(3):863–869
7. Altuntas E, Weber C, Schubert US (2013) Detailed characterization of poly(2-ethyl-2-oxazoline)s by energy variable collision-induced dissociation study. Rapid Commun Mass Spectrom 27(10):1095–1100

Chapter 6

Heterofunctional Glycopolypeptides by Combination of Thiol-Ene Chemistry and NCA Polymerization

Kai-Steffen Krannig and Helmut Schlaad

Abstract

Glycopolypeptides are prepared either by the polymerization of glycosylated amino acid *N*-carboxyanhydrides (NCAs) or by the post-polymerization functionalization of polypeptides with suitable functional groups. Here we present a method for the in-situ functionalization and (co-) polymerization of allylglycine *N*-carboxyanhydride in a facile one-pot procedure, combining radical thiol-ene photochemistry and nucleophilic ring-opening polymerization techniques, to yield well-defined heterofunctional glycopolypeptides.

Key words Amino acid *N*-carboxyanhydride, Glyco, Photochemistry, Polypeptides, Thiol-ene

1 Introduction

Synthetic polypeptides, and especially glycopolypeptides, recently attracted increasing interest as promising materials for applications in biomedicine and biotechnology, e.g., tissue engineering, drug delivery, or as polymer therapeutics [1–4]. The synthesis, structural characteristics, self-assembly behavior, and ability of glycopolypeptides to recognize and selectively bind to proteins (lectins) have been investigated and highlighted in numerous articles, manifesting the increasing importance of this class of materials [2, 5–9]. Although significant progress has been made [8], the preparation of well-defined glycopolypeptides is still a challenging task for synthetic polymer chemists. Recent efforts include the polymerization of glycosylated NCAs as well as the post-polymerization functionalization of ready-made polypeptides carrying appropriate functional groups in the side chains. However, most of these approaches require multiple steps and sophisticated, tedious purification protocols. Especially the works with hydrolytically unstable NCAs require careful and skillful handling and are labor- and time-consuming processes.

Xue-Long Sun (ed.), *Macro-Glycoligands: Methods and Protocols*, Methods in Molecular Biology, vol. 1367, DOI 10.1007/978-1-4939-3130-9_6,

We suggested a very facilitated (and transition metal-free) synthesis of glycopolypeptides by in situ glycosylation and polymerization of AGly NCA, combining radical thiol-ene photochemistry and nucleophilic ROP (Fig. 1a) [10]. The so prepared glycopolypeptides contain a predetermined amount of sugar and remaining vinyl groups, which in a second step can be functionalized to yield heterofunctional glycopolypeptides with a variety of functionalities (exemplary here: carboxyl and glucosyl) (Fig. 1b). These additional functionalities (e.g., amine, ethylene glycols, other carbohydrates) could be used to introduce stimuli-responsiveness or trigger the folding of polypeptide chains into higher order structures.

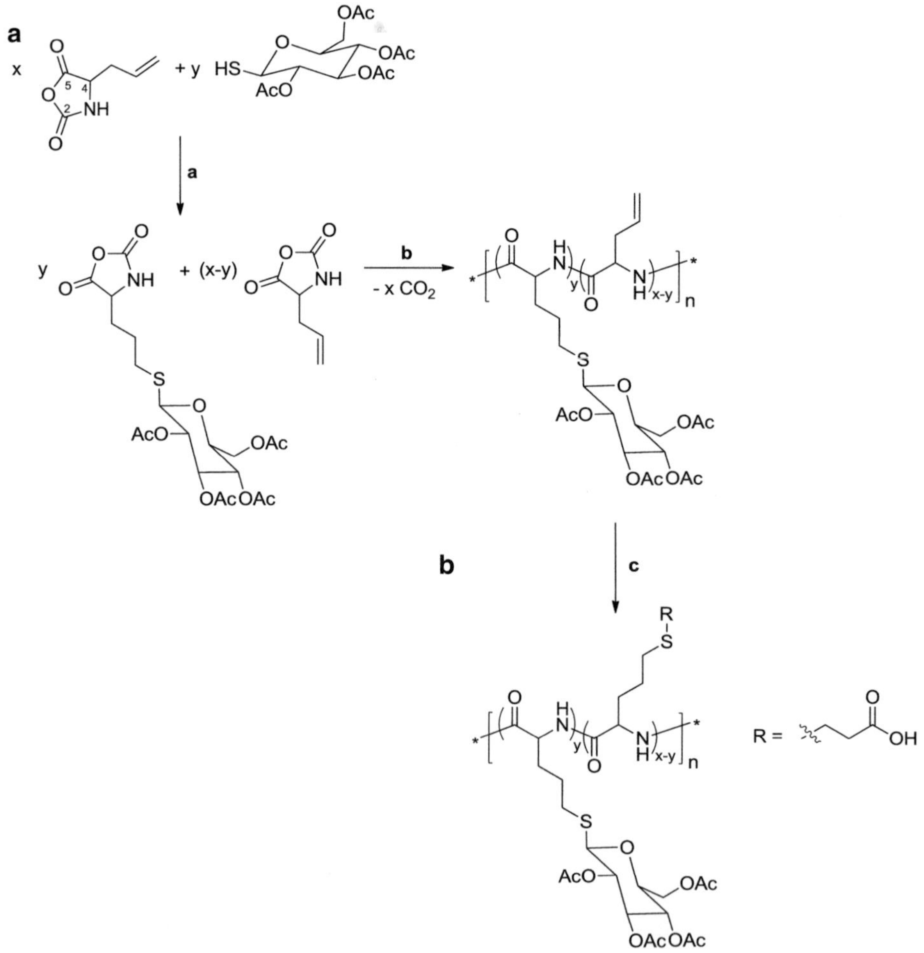

Fig. 1 (**a**) One-pot partial glycosylation and copolymerization of AGly NCA and (**b**) subsequent functionalization with thiol to yield heterofunctional glycopolypeptides. Reagents and conditions: *a* thiol-ene photoaddition: benzophenone, hν, THF, r.t., 45 min ($y < x$); *b* nucleophilic ROP: 1-hexylamine, THF/DMF, r.t., 7 days; *c* 3-mercaptopropionic acid, benzophenone, hν, THF, r.t., 16 h. Adapted from [10]

2 Materials

2.1 Chemicals

1. DL-Allylglycine (>98 %) (BoaoPharma).
2. Benzophenone (Sigma-Aldrich).
3. *N,N*-Dimethylformamide (≥99.8 %, extra dry) (Sigma-Aldrich).
4. Ethyl acetate (Th. Geyer GmbH & Co, KG), dried over CaH_2 and distilled.
5. Heptanes (99 %) *n*-heptane (99 %) (Roth).
6. 1-Hexylamine (>99.5 %) (Sigma-Aldrich).
7. Isopropanol (tech.)
8. 3-Mercaptopropionic acid (99 %+) (Sigma-Aldrich).
9. α-Pinene (Alfa Aesar).
10. Silica-gel (Fluka), dried at 150 °C for 48 h
11. Tetrahydrofuran (99.5 %, extra dry), 1,4-dioxane (Acros Organics).
12. 1-Thio-β-D-glucose-2,3,4,6-tetraacetate (97 %) (Sigma-Aldrich).
13. Triphosgene (Merck).

2.2 UV Light Source

1. Energy saving lamp Exo Terra ReptiGlo 5.0, 26W (Fig. 2) just below Subheading 2.2 UV light source.

3 Methods

3.1 Monomer Synthesis [11]

1. Add (suspend) 2.5 g of AGly (21.7 mmol, 1.0 equiv) in 100 mL of THF and heat to 50 °C.

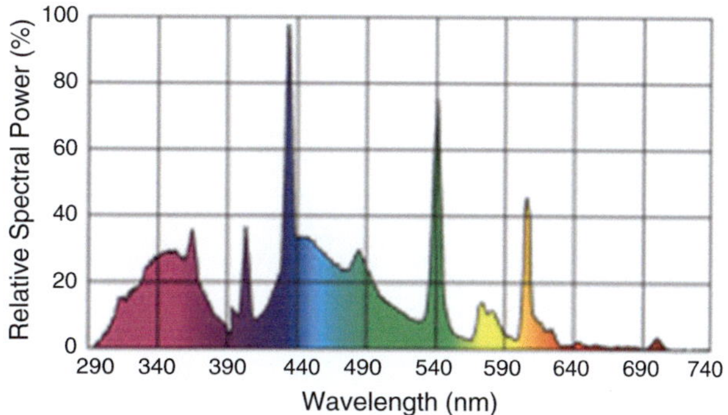

Fig. 2 Energy saving lamp Exo Terra ReptiGlo 5.0, 26W (*left*) and UV/vis emission spectrum (*right*) (as provided by the manufacturer). Reprinted with permission from [11]. Copyright 2012, American Chemical Society

2. At this temperature, add 13.75 mL of α-pinene (86.8 mmol, 4.0 equiv) and 2.57 g of triphosgene (8.7 mmol, 0.4 equiv) and flush a constant stream of argon through the reaction mixture.
3. A clear solution usually forms within 45 min; otherwise add additional triphosgene (0.05 equiv/30 min).
4. After 3 h, concentrate the solution to 1/3 of the volume and precipitate in excess heptanes.
5. Collect the white precipitate and remove residual heptanes under high vacuo (ca. 1 h).
6. Redissolve the powder in minimum amount of ethyl acetate and filter through standard filter paper into tenfold volume of heptanes.
7. Repeat **steps 5** and **6** two times.
8. Collect the white precipitate and remove residual solvent under high vacuo. Yield: 1.6 g (11.3 mmol, 52 %) (*see* **Note 1**)
9. Characterize the product by melting point (*see* **Note 2**) and ^{1}H NMR (*see* **Note 3**). Melting point: 89–90 °C. ^{1}H NMR (400 MHz, $CDCl_3$): δ (ppm) = 6.59 (s, 1H, N*H*), 5.74 (m, 1H, $H_2C{=}CH$), 5.28 (m, 2H, $H_2C{=}CH$), 4.40 (dd, ^3J = 7.0 Hz, ^4J = 4.3 Hz, 1H, H_2C-C*H*-NH), 2.53 (td, ^3J = 14.6 Hz, ^3J = 7.4 Hz, 1H diast., H_2C-CH-NH), 2.10 (td, ^3J = 14.6 Hz, ^3J = 7.4 Hz, 1H diast., H_2C-CH-NH).

3.2 One-Pot Glycosylation/ Polymerization (Exemplary Procedure)

1. Dissolve AGly NCA (1.0 equiv), benzophenone (0.2 eq), and the respective amount of AcGlcSH (0.8 equiv) in dry THF (0.15 M) under an argon atmosphere.
2. Irradiate the reaction mixture with UV light from two energy saving lamps (Exo Terra ReptiGlo 5.0 26W) (distance UV lamp to reaction vessel: ca. 5 cm) for ~45 min (*see* **Note 4**).
3. Remove vessel from the lamps and add dry DMF (overall concentration 5 wt%) and desired amount of a 0.1 M solution of freshly distilled 1-hexylamine ($[NCA]_0/[amine]_0 = 30$) in dry DMF
4. Stir the reaction mixture for 7 days under reduced pressure (ca. 0.5 mbar) at room temperature (*see* **Note 5**).
5. Quench the polymerization by precipitation into a tenfold volume of isopropanol.
6. Collect the product by centrifugation and dry at 65 °C in high vacuum. Isolated yield: 80 %.
7. Characterize the product by ^{1}H NMR (*see* **Note 3**) and SEC (*see* **Note 6**) and ^{1}H NMR (400 MHz, TFA-d): δ (ppm) = 5.6–5.8 (-*H*C=C-), 5.6–5.4 (S-C*H*-O), 5.4–5.3 (Glc), 5.3–5.2 (-HC=CH_2, Glc), 4.3–4.8 (C(=O)-C*H*-NH), 3.8–4.0 (Glc),

3.4 (CH_2-$\mathit{CH_2}$-NH_2), 2.9–2.4 (S-$\mathit{CH_2}$), 2.3–1.6 (S-CH_2-CH2-$\mathit{CH_2}$, O*Ac*), 1.3–1.2 (CH_3-$\mathit{CH_2}$-$\mathit{CH_2}$-$\mathit{CH_2}$-$\mathit{CH_2}$-), 0.8 ($\mathit{CH_3}$). Composition (AGly)/(GlcAGly) = 0.32/0.68 (^{1}H NMR), average number of Gly repeat units: 28 (^{1}H NMR end group analysis), number-average molar mass: M_n^{app} = 8500 g/mol (SEC), dispersity: $Đ$ = 1.22 (SEC).

3.3 Post-polymerization Functionalization (Exemplary Procedure)

1. Dissolve partially glycosylated polypeptide ((AGyl)/(GlcAGly) = 0.32/0.68 (9/19 units)), benzophenone (0.1 equiv with respect to double bonds), and 3-mercaptopropionic acid (1.5 equiv with respect to double bonds) in THF (ca. 1.0 wt% with respect to AGly units) and put it under an inert argon atmosphere.
2. Seal the vessel and irradiate it with UV light for 16 h (*see* **Note 7**).
3. Dilute the reaction mixture and extensively dialyze (RC 1000) against THF (*see* **Note 8**).
4. Removal of THF and freeze-drying from 1,4-dioxane yield the final products as fluffy solids.
5. ^{1}H NMR (400 MHz, TFA-d): Fig. 3. Quantitative conversion of AGly units (^{1}H NMR), number-average molar mass: M_n^{app} = 10,280 g/mol (SEC), dispersity: $Đ$ = 1.26 (SEC).

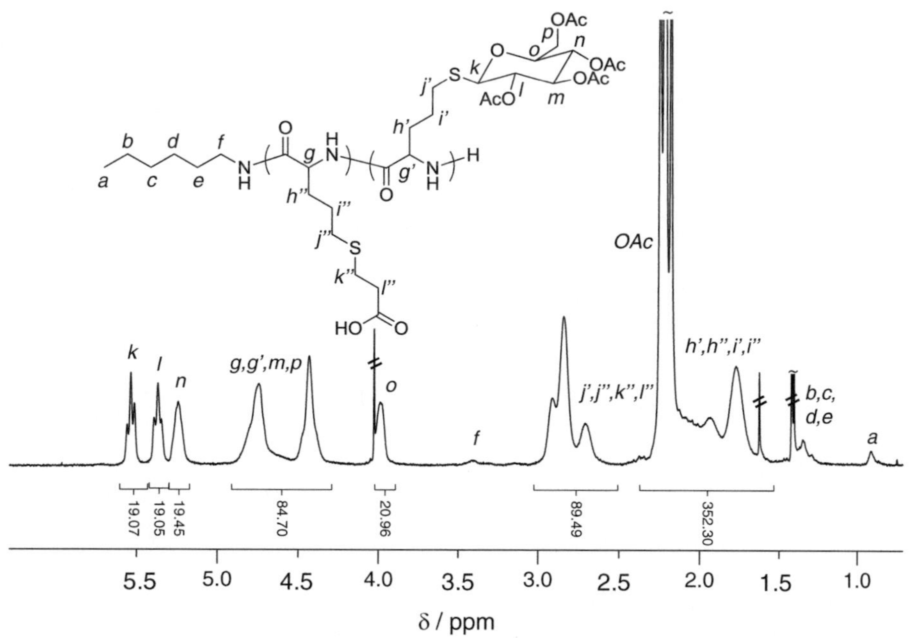

Fig. 3 ^{1}H-NMR spectrum (400 MHz, TFA-d) of carboxylated glycopolypeptide. Reprinted with permission from [10]. Copyright 2014, American Chemical Society

4 Notes

1. The AGly NCA monomer needs to be stored in the freezer and is stable for up to 3 months.
2. Melting points are determined using a MEL-TEMP® apparatus from Lab Devices INC, USA with a Fluke 51 thermometer.
3. ^{1}H NMR measurements are conducted at room temperature using a Bruker DPX-400 spectrometer operating at 400 MHz. Deuterated chloroform and TFA are used as solvents (Sigma-Aldrich); ^{1}H NMR signals are referenced to the signals of $CDCl_3$ δ 7.26 ppm and TFA-d δ 11.52 ppm, respectively.
4. Irradiation must start immediately after mixing of the reactants. The reaction can be accelerated by using more lamps.
5. The impact of pressure on polymerization has not been investigated. Key is the removal of CO_2 which is released during monomer addition.
6. SEC with simultaneous UV and RI detection is performed with NMP (+0.5 wt% LiBr) as the eluent, flow rate: 0.8 ml/min, at 70 °C using a set of two 300 × 8 mm^2 PSS-GRAM columns with average particle sizes of 7 μm and porosities of 100 and 1000 Å. Calibration was done using poly(methyl methacrylate) standards (PSS, Mainz, Germany).
7. Full conversion of monomer is usually achieved within 3–5 h.
8. Dialysis bags are becoming brittle in THF and should be handled with care (to avoid damage or rupture).

Acknowledgment

Financial support was given by the Max Planck Society and the German Research Foundation (within the IUPAC Transnational Pilot Call in Polymer Chemistry).

References

1. Deming TJ (2007) Synthetic polypeptides for biomedical applications. Prog Polym Sci 32:858–875
2. Huang J, Heise A (2013) Stimuli responsive synthetic polypeptides derived from N-carboxyanhydride (NCA) polymerisation. Chem Soc Rev 242:7373–7390
3. Duncan R (2003) The dawning era of polymer therapeutics. Nat Rev Drug Discov 2:347–360
4. Ringsdorf H, Schlarb B, Venzmer J (1988) Molecular architecture and function of polymeric oriented systems: models for the study of organization, surface recognition, and dynamics of biomembranes. Angew Chem Int Ed Engl 27:113–158
5. Bonduelle C, Lecommandoux S (2013) Synthetic glycopolypeptides as biomimetic analogues of natural glycoproteins. Biomacromolecules 14:2973–2983

6. Quadir MA, Martin M, Hammond PT (2014) Clickable synthetic polypeptides: routes to new highly adaptive biomaterials. Chem Mater 26:461–476
7. Kricheldorf HR (2006) Polypeptides and 100 years of chemistry of alpha-amino acid *N*-carboxyanhydrides. Angew Chem Int Ed 45:5752–5784
8. Kramer JR, Deming TJ (2014) Recent advances in glycopolypeptide synthesis. Polym Chem 5:671–682
9. Krannig K-S, Schlaad H (2014) Emerging bio-inspired polymers: glycopolypeptides. Soft Matter 10:4228–4235
10. Krannig K-S, Doriti A, Schlaad H (2014) Facilitated synthesis of heterofunctional glycopolypeptides. Macromolecules 47:2536–2539
11. Krannig K-S, Schlaad H (2012) pH-responsive bioactive glycopolypeptides with enhanced helicity and solubility in aqueous solution. J Am Chem Soc 134:18542–18545

Chapter 7

Preparation of Proteoglycan Mimetic Graft Copolymers

Matt J. Kipper and Laura W. Place

Abstract

Proteoglycans are proteins with pendant glycosaminoglycan polysaccharide side chains. The method described here enables the preparation of graft copolymers with glycosaminoglycan side chains, which mimic the structure and composition of proteoglycans. By controlling the stoichiometry, graft copolymers can be obtained with a wide range of glycosaminoglycan side-chain densities. The method presented here uses a three-step reaction mechanism to first functionalize a hyaluronic acid backbone, followed by reductive amination to couple the glycosaminoglycan side chain to the backbone, by the reducing end. Proteoglycan mimics like the ones proposed here could be used to study the structure–property relationships of proteoglycans and to introduce the biochemical and biomechanical properties of proteoglycans into biomaterials and therapeutic formulations.

Key words Proteoglycans, Graft copolymers, Glycosaminoglycans, Heparin, Chondroitin sulfate, Hyaluronan, Aggrecan, Versican

1 Introduction

1.1 Proteoglycans

Proteoglycans (PGs) are an important class of proteins modified with glycosaminoglycan (GAG) side chains (Fig. 1) [1]. The GAG side chains are covalently attached at their reducing ends to the core protein by an *O*-linked tetrasaccharide bound to a serine residue in the protein. Like all proteins, the sequence of the PG core is predetermined, according to a genetic blueprint. However, the GAG side chains are synthesized without a template; as sugar residues are added by glycosyl transferases, other enzymes, such as epimerases and sulfotransferases, modify sugars in the growing GAG. The activities and substrate specificities of the enzymes involved in GAG synthesis and modification may vary by organism, tissue type, and the evolving sequence of the GAG. The resulting PGs can thereby have very specific functionality provided by the core protein, and diverse biochemical functions provided by the length and sequence variability of the GAG side chains.

Xue-Long Sun (ed.), *Macro-Glycoligands: Methods and Protocols*, Methods in Molecular Biology, vol. 1367,
DOI 10.1007/978-1-4939-3130-9_7, © Springer Science+Business Media New York 2016

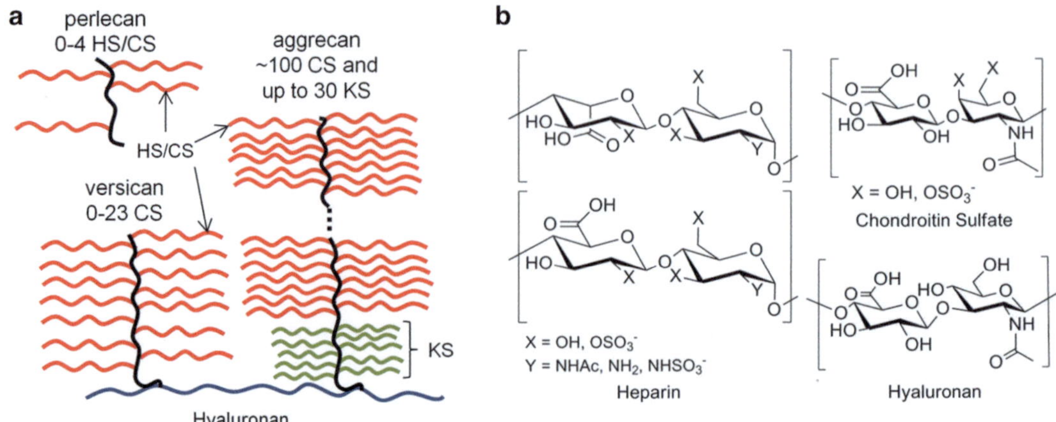

Fig. 1 (**a**) Large proteoglycan structures. Protein cores are *black*, HS and CS are *red*, keratan sulfate (KS) is *green*, and hyaluronan is *blue*. (**b**) Structures of the GAGs used in this work

These GAGs comprising the PG side chains include chondroitin sulfate (CS), dermatan sulfate (DS), keratan sulfate (KS), heparin, and heparan sulfate (HS). These GAGs all have an alternating sequence of a hexuronic acid (or hexose, in the case of DS) and an acetylated hexosamine, and are classified according to their constituent disaccharides and glycosidic links. Within each type of GAG, sequence variability arises from different modifications of the saccharide residues. These modifications primarily include *O*-sulfation on the 4-carbon or the 6-carbon of the hexosamine and the 2-carbon or 6-carbon of the hexuronic acid (or hexose), epimerization of the hexuronic acid, and 2-*N*-deacetylation and 2-*N*-sulfonation of the hexosamine. Combinations of these modifications give rise to a wide variety of possible saccharide sequences.

1.2 Proteoglycan Organization

PGs are found associated with the cell surface (e.g., syndecans and glypicans) with their glycosylated regions extending into the pericellular space, or they are secreted (e.g., versican and aggrecan) and contribute important functions to the extracellular matrix (ECM) [2–4]. Some PGs, such as serglycin are intracellular, but even serglycin is associated with proteases, cytokines, and secretory vesicles that ultimately have extracellular functions [5–7]. Therefore, the most well-characterized functions of PGs are associated with extracellular and pericellular phenomena, such as cell adhesion, cytokine signaling, organization of ECM components, and maintenance of physicochemical, transport, and mechanical properties of the ECM.

PGs can be organized into larger functional assemblies, such as the glycocalyx on the surfaces of cells and the aggrecan aggregate in the ECM. The glycocalyx is a GAG-rich region found on the surface of many eukaryotic and prokaryotic cells. The pericellular region defined by the glycocalyx is the immediate context in which cells interact with everything in their environment. So understanding its structure–function relationships is critical for understanding many biological phenomena. For example, in the vascular endothelium the glycocalyx regulates interactions between blood and the vessel wall. In the ECM of skeletal tissues the GAG hyaluronan binds to proteoglycans forming GAG-rich assemblies with critical biomechanical and biochemical functions.

Secreted proteoglycans vary widely in the number of GAG side chains, from 0 to 4 HS or CS side chains (perlecan), up to 23 CS chains (versican), to over 100 CS and KS chains (aggrecan) (Fig. 1) [3, 8–11]. Versican and aggrecan both have regions that bind to hyaluronan in the ECM, enabling their further assembly into higher order aggregates. The resulting high density of negatively charged sulfate groups binds counterions, maintains high osmotic pressure, and thereby regulates the hydration and mechanical properties of tissues [7, 19, 22]. The high density of GAGs in these assemblies also may provide a reservoir of bound cytokines, which are stabilized in the ECM by GAG-protein interactions.

1.3 Proteoglycan Mimics

The many biological functions of GAGs are in part regulated by their organization, and one of the primary functions of PGs is to organize GAGs at cell surfaces and in the ECM. To better understand how this organization influences the biological functions of GAGs, several approaches have been recently proposed to assemble GAGs into PG mimics [12, 13]. By tuning their structure and composition, researchers might better understand the structure–function relationships of PGs, and how these relationships can be exploited to design functional GAG-based nanomaterials.

Recent approaches to developing proteoglycan mimics have included methods for preparing GAG-containing nanoparticles and modifying surfaces with GAGs. Recently, the Panitch group has prepared GAG-modified peptides that bind the ECM components collagen and hyaluronan [14–18]. These have on average one GAG chain per peptide, and may mimic the assembly of PGs in the ECM, and their ECM-protective properties. The Jiang group has recently prepared vesicles containing glycopolymers with pendant glucose or galactose residues. These present a high density of saccharide groups to mimic the glycocalyx, though they contain neutral glycosides [19]. GAG-polymer complexes, like those recently reported by the Marcolongo group and the Hsieh-Wilson group, are particularly useful for surface modification with a high density of GAG chains and oligosaccharides [13, 20]. Our research group has reported polyelectrolyte complexes containing GAGs that can also be used to either modify surfaces or be delivered to cells in culture in solution. These have the ability to bind and deliver growth factors, mimicking the behavior of the GAG aggrecan [21–24].

Here we detail the methods we have recently reported for preparing graft copolymers that have GAG side chains [25]. This method has the advantage that the GAG content of the graft copolymers can be tuned over a broad range to obtain graft copolymers with different degrees of GAG substitution to mimic a broad range of PGs, like those shown in Fig. 1. They can also be adsorbed to surfaces to present a high density of GAGs at surfaces. The graft copolymers are composed of a polymer backbone (modified hyaluronan) with pendent functional (azide) groups that are reacted with the reducing end of GAGs to covalently graft the side chains. This is done using a series of three reactions described below.

Reaction 1: The hyaluronan backbone is first functionalized with thiols.

Carboxylic acids along the hyaluronan chain are coupled with cysteamine using EDC/NHS chemistry.

The thiolated hyaluronan is then purified by dialysis.

The intermediate is lyophilized and collected for use in the second reaction.

Reaction 2: The thiolated hyaluronan is activated with a hydrazide.

To maximize reactivity, the thiols along the backbone are reduced to break any disulfide bonds and purified using spin columns.

An Ellman's reagent assay can be done to quantify active sites for coupling.

Thiols are then reacted with a coupling agent that is maleimide-activated on one end and hydrazide-activated on the other.

The hydrazide hyaluronan is then purified by dialysis.

The intermediate is lyophilized and collected for use in the third reaction.

Reaction 3: GAGs are grafted onto the hyaluronan backbone via reductive amination.

The hydrazide groups on the hyaluronan backbone react with the reducing end of the GAG. A strong reducing agent is added intermittently throughout the reaction.

The solvent is removed and the product is redissolved in water and dialyzed for purification.

The final product is lyophilized and collected.

The neat polymers, intermediates, and final products are characterized using FT-IR, NMR, and DLS.

2 Materials

2.1 Major Equipment/Instruments

1. pH meter.
2. Oil bath.
3. Heated/stir plate with thermocouple.
4. Plate reader.
5. Lyophilizer.
6. Centrifuge with adapter for 50 mL centrifuge tubes.
7. Nuclear magnetic resonance (NMR) spectrometer.
8. Fourier-transform infrared spectrometer.
9. Dynamic light scattering instrument.
10. Zeta potential Instrument.
11. 0.22 μm sterile poly(ethersulfone) (PES) or cellulose nitrite (CN) filter units.
12. 0.22 μm poly(vinylidene fluoride) (PVDF) syringe filters.

2.2 Chemicals for Backbone Thiolation

1. Sodium hyaluronate (HA; $M_w = 740$ kDa) (*see* **Notes 1** and **2**).
2. MES buffer (2-(*N*-morpholino)ethanesulfonic acid (MES) sodium salt, 0.5 M sodium chloride in diH_2O, and 1 M sodium hydroxide in diH_2O to adjust pH to 6.0).
3. *N*-(3-dimethylaminopropyl)-*N*′-ethylcarbodiimide hydrochloride (EDC).
4. *N*-hydroxysuccinimide (NHS).
5. Cysteamine hydrochloride.
6. Dialysis buffers (0.5, 0.25, 0.1, 0.05 M sodium chloride in diH_2O).
7. Seamless cellulose dialysis tubing (12 kDa MWCO).
8. Ultrapure water (diH_2O; 18.2 MΩ cm).

2.3 Chemicals for Backbone Hydrazide Activation

1. Thiolated hyaluronan backbone (HA-SH).
2. Phosphate buffered saline (0.1 M sodium phosphate dibasic and 0.15 M sodium chloride in diH_2O and 0.1 M sodium phosphate monobasic and 0.15 M sodium chloride in diH_2O).
3. Tris(2-carboxyethyl)phosphine hydrochloride (TCEP-HCl).
4. 5,5′-Dithio-bis-[2-nitrobenzoic acid] (Ellman's reagent).
5. Ellman's reagent assay reaction buffer (0.1 M sodium phosphate dibasic, 0.15 M sodium chloride, 1 mM ethylenediaminetetraacetic acid (EDTA) in diH_2O and 0.1 M sodium phosphate monobasic and 0.15 M sodium chloride in diH_2O to adjust pH to 8).

6. Cysteine hydrochloride monohydrate.
7. *N*-[β-maleimidopropionic acid] hydrazide trifluoroacetic acid salt (BMPH) (*see* **Note 3**).
8. Zeba Spin Desalting Columns (10 mL, 7 kDa MWCO) (*see* **Note 4**).
9. 50 mL centrifuge tubes.
10. Dialysis buffers (0.5, 0.25, 0.1, 0.05 M sodium chloride in diH_2O).
11. Seamless cellulose dialysis tubing (12 kDa MWCO).
12. Ultrapure water (diH_2O; 18.2 MΩ cm).

2.4 Chemicals for Coupling via Reductive Amination

1. Hydrazide activated hyaluronan backbone (HA-BMPH).
2. Chondroitin sulfate sodium salt (CS; from shark cartilage, 6 % sulfur, 6 sulfate/4 sulfate = 1.24, M_w = 84.3 kDa) (*see* **Note 5**).
3. Heparin sodium (from porcine intestinal mucosa, 12.5 % sulfur, Mw = 14.4 kDa) (*see* **Note 5**).
4. *N,N*-dimethylformamide (DMF).
5. Sodium triacetoxyborohydride (STAB).
6. Glacial acetic acid.
7. Dialysis buffers (0.5, 0.25, 0.1, 0.05 M sodium chloride in diH_2O).
8. Biotech cellulose ester membrane dialysis tubing (300 kDa MWCO) (*see* **Note 6**).
9. Ultrapure water (diH_2O; 18.2 MΩ cm).

3 Methods

3.1 Methods for Hyaluronan Backbone Thiolation

1. Dissolve 0.1 M MES and 0.5 M NaCl in diH_2O.
2. Dissolve 1 M sodium hydroxide in diH_2O.
3. Adjust the pH of the MES solution to 6.0 by titrating with the sodium hydroxide solution.
4. Filter the buffer with 0.22 μm membrane unit.
5. Dissolve 250 mg of HA (*see* **Notes 7** and **8**) in 50 mL of MES buffer solution, heated to 45 °C in an oil bath, overnight (*see* **Note 9**). The oil bath setup is shown in Fig. 2.
6. Remove the solution from the oil bath and allow it to cool to room temperature.
7. Add 0.645 g of EDC-HCl (10× molar excess relative to carboxylate functional groups on HA) and 0.967 g of NHS (*see* **Note 10**). Stir using a stir bar and stir plate, as instructed in

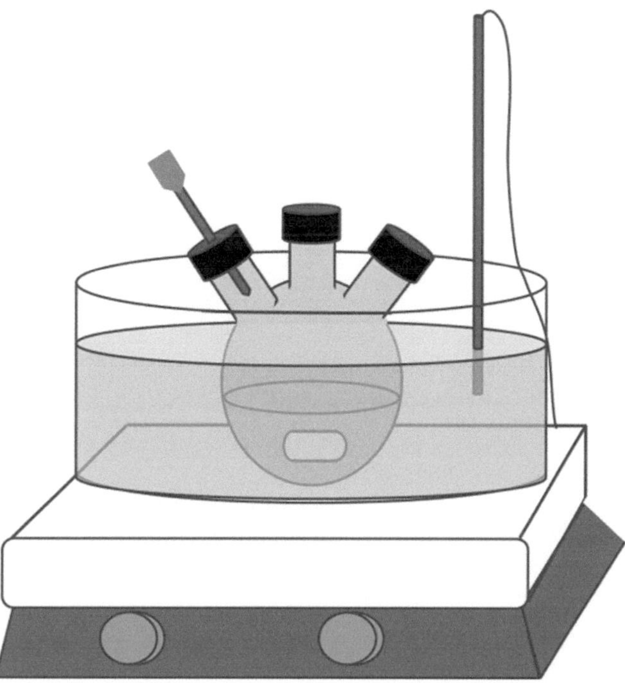

Fig. 2 Dissolve HA in MES buffer solution overnight using a heated oil bath at 45 °C

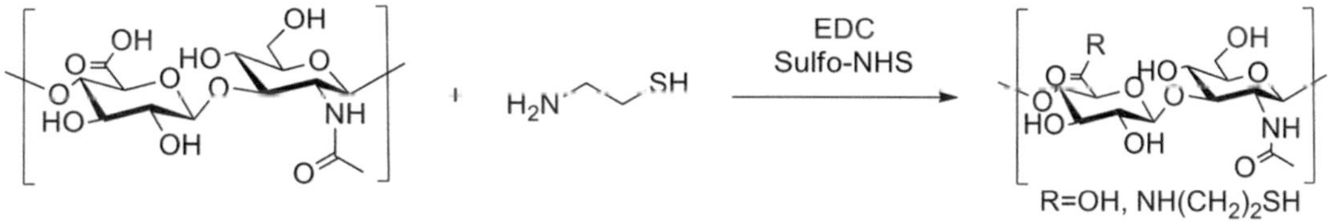

EDC
Sulfo-NHS

R=OH, NH$(CH_2)_2$SH

Fig. 3 Synthesis of HA-SH intermediate

Note 7, and react for 2 h to activate the carboxylate functional group on HA (Fig. 3).

8. Raise the pH to 7.2 using sodium hydroxide (*see* **Note 11**).
9. Add 0.88 g of cysteamine hydrochloride (23× molar excess compared to HA starting material) to the reaction vessel and stir for 5 h at room temperature.
10. Using a pipette, transfer the resulting solution into 12 kDa MWCO dialysis tubing and place the tubing into a large beaker filled with 0.5 M sodium chloride in diH_2O (*see* **Note 12**). During dialysis, stir with a stir bar and stir plate very slowly (*see* Fig. 4).
11. Dialyze the product for 5 days against decreasing concentrations of sodium chloride (0.5, 0.25, 0.1, 0.05 M sodium chloride, and finally against diH_2O) changing dialysis buffer every

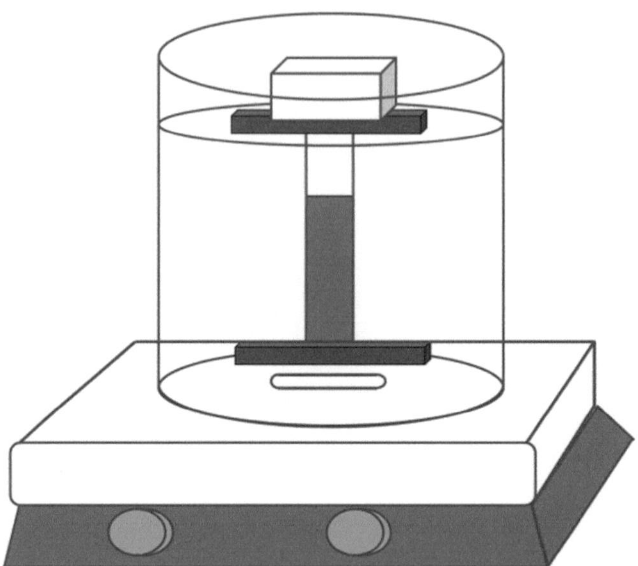

Fig. 4 Dialyze the HA-SH to remove unreacted cysteamine

24 h. This removes excess cysteamine hydrochloride, EDC, and salt (*see* **Note 13**).

12. After dialysis, lyophilize to recover a solid product (thiolated HA; HA-SH). To do this, first freeze the solution completely and then attach to a lyophilizer with the vacuum pump running. This generally takes 24–48 h depending upon the starting volume (*see* **Note 14**).
13. Recover the product and store at 4 °C.

3.2 Methods for Backbone Hydrazide Activation

1. Dissolve 0.1 M sodium phosphate dibasic and 0.15 M sodium chloride in diH_2O.
2. Dissolve 0.1 M sodium phosphate monobasic and 0.15 M sodium chloride in diH_2O.
3. Prepare a pH 8 and a pH 7.2 phosphate buffer by combining these two solutions.
4. Dissolve 100 mg of HA-SH in 50 mL of the pH 8 phosphate buffer.
5. Dissolve 114 mg of TCEP (3× molar excess compared to thiol groups) in the above solution and allow the reaction to continue for 1 h at room temperature. This reduces any disulfide bonds that may have formed between thiol groups on the HA-SH so that they are available for the following reaction (*see* **Note 15**).

6. To remove excess TCEP and to exchange the pH 8 buffer with pH 7.2 buffer, pass the solution through desalting columns.
7. First, the desalting column must be equilibrated. Twist off the bottom closure and loosen the cap of the spin column and place in a 50 mL centrifuge tube to collect storage buffer. Centrifuge this assembly at 1000 × *g* for 2 min (*see* **Notes 16** and **17**). Discard the storage buffer from the 50 mL collection tube.
8. Mark the side of the spin column where the resin is slanted upwards using a permanent marker. Be sure to arrange the spin column with this mark facing out in all subsequent centrifugation steps.
9. Replace the spin column into the collection tube. Add 5 mL of pH 7.2 PBS to the top of the spin column gently, so as not to disturb the resin bed. Centrifuge at 1000 × *g* for 2 min. Discard buffer in collection tube. Repeat this two or three times to fully equilibrate spin column.
10. Place the spin column in a new clean collection tube. Gently add 5 mL of sample to the top of the resin bed. Centrifuge at 1000 × *g* for 2 min. Collect the sample from the collection tube; this contains reduced HA-SH in PBS pH 7.2. *See* Fig. 5 for spin column graphic.
11. Combine all of the collected samples into a round bottom flask and add 80 mg of BMPH (2× molar excess compared to SH groups on the hyaluronan backbone) to the solution. Place a stir bar in the flask and stoppers in the neck(s) of the flask and

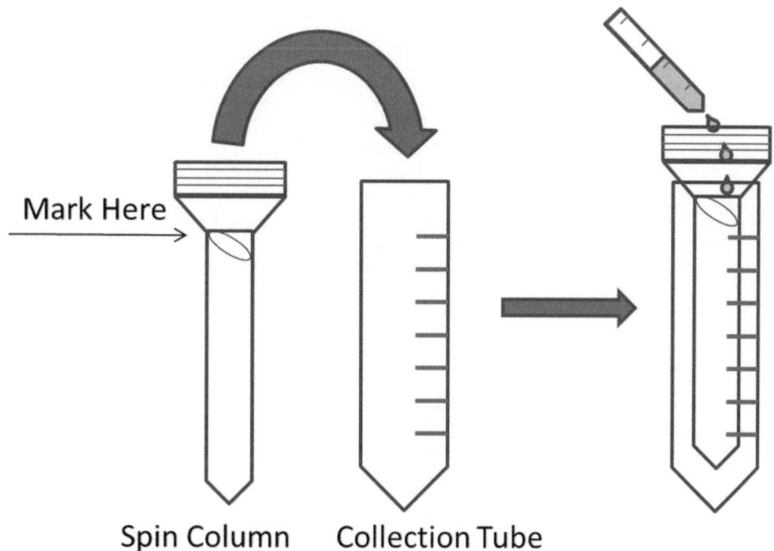

Fig. 5 Mark and load the spin column into a 50 mL centrifuge tube

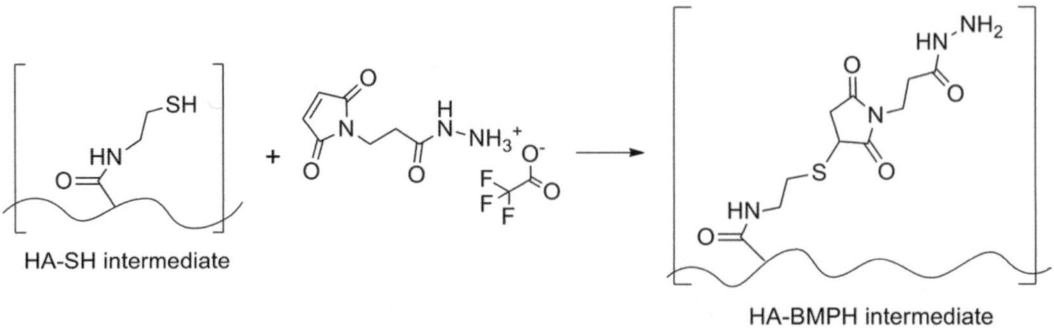

Fig. 6 Synthesis of HA-BMPH intermediate

stir using a stir plate. Allow the coupling reaction to proceed for 2 h at room temperature (Fig. 6).

12. Dialyze the solution as described in **step 11** of Subheading 3.1 to remove unreacted BMPH.
13. Lyophilize as described in **step 12** of Subheading 3.1 to obtain solid HA-BMPH intermediate.

3.3 Methods for Coupling via Reductive Amination

1. Place HA-BMPH (15 mg) and either CS (1 g) or heparin (170 mg) and a stir bar into a round-bottom flask (*see* **Note 18**).
2. Seal the flask, add 10 mL of anhydrous DMF, and purge with nitrogen (*see* **Note 19**).
3. Add 350 μL of acetic acid to the vessel as shown in Fig. 7.
4. Place the flask in an oil bath with a thermocouple and turn on the stir plate as described in **Note** 7. Heat the reaction stepwise from 25 to 85 °C by increasing the temperature by 10 °C every 20 min.
5. In a separate round-bottom flask, dissolve 1 g of STAB (100× molar excess) in 10 mL of anhydrous DMF and purge with nitrogen.
6. Once the reaction flask has reached 85 °C, add 350 μL of STAB, as shown in Fig. 5.
7. Add 350 μL of STAB every 2 h four more times (for a total of 10 h and 1.75 mL of STAB) and allow the reaction to continue overnight. The reaction is illustrated in Fig. 8.
8. After the reaction the final product must be purified.
9. Transfer the reaction mixture to a round-bottom purification flask with a stir bar and seal the top. Attach vacuum tubing to the arm of the flask and attach the other end to a solvent trap. Attach another piece of vacuum tubing to the arm of the solvent trap and attach the final end to a vacuum pump. Place the flask in an oil bath (set at 60 °C) and turn on the stir plate. Place the solvent trap in a Dewar flask filled with either liquid

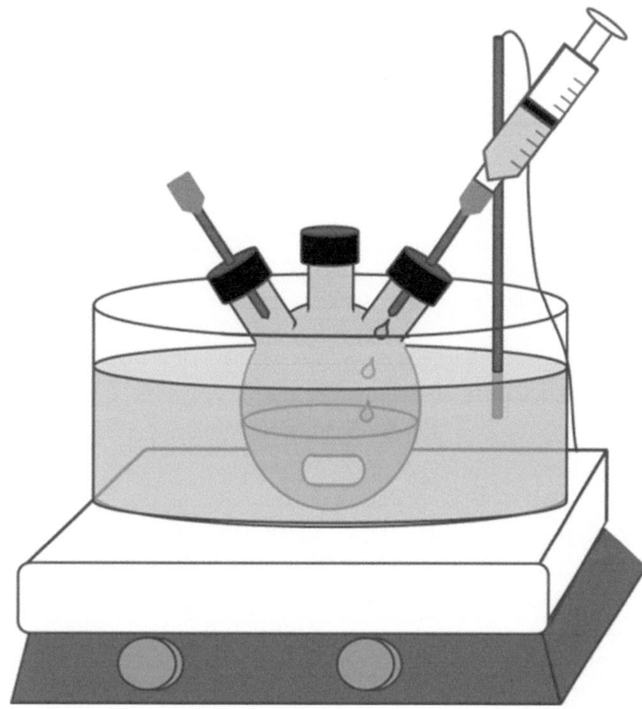

Fig. 7 Add reagents to the reaction vessel

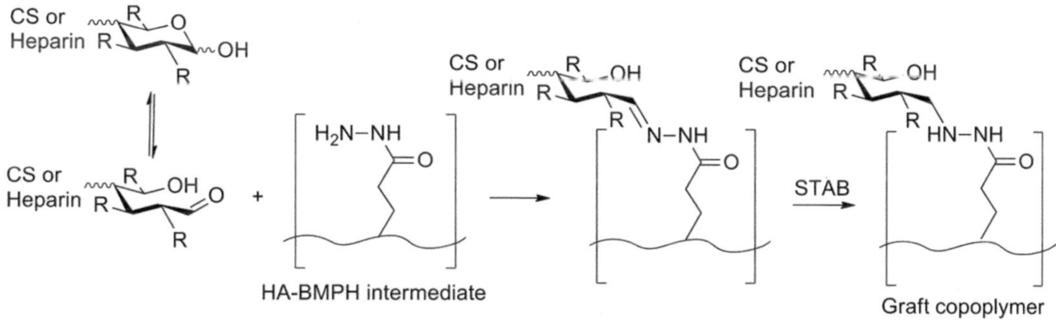

Fig. 8 Coupling CS or heparin to HA-BMPH to form graft copolymers

nitrogen or dry ice and isopropanol (*see* **Note 20**). Close the cap on the purification flask, turn on the vacuum pump, and slowly open the cap until boiling is apparent in the flask. This apparatus is shown in Fig. 9.

10. Once the DMF has been removed, dissolve the product in as little diH_2O as possible (*see* **Note 21**).
11. Transfer the product into dialysis tubing (300 kDa) and dialyze as described in **step 11** of Subheading 3.1. Dialysis using large molecular weight cutoff is necessary to remove unreacted GAGs.

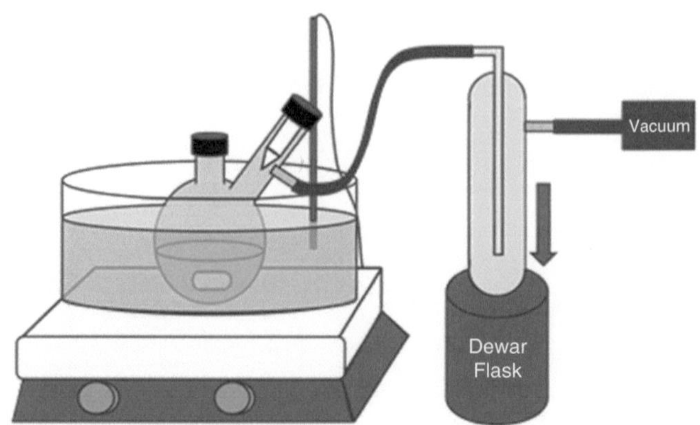

Fig. 9 Apparatus for removing DMF from the reaction mixture

12. Lyophilize the sample, recover the final product, and store at 4 °C.

3.4 Methods for Chemical Characterization of PG Mimics

1. Use attenuated total reflectance Fourier-transform infrared spectroscopy (ATR-FTIR) to characterize the neat polymers, the intermediates, and the final product (*see* **Notes 22** and **23**).
2. Deposit dry powder of the product (or the polymer to be analyzed) onto a ZnSe crystal and collect spectra from 4000 to 650 cm^{-1}. Absorption bands of the proteoglycan mimics should be characteristic of those observed in the neat GAGs [25]. These absorptions include broad OH stretching near 3280 cm^{-1}, several carbonyl-stretching bands between 1600 and 1700 cm^{-1}, NH bending near 1560 cm^{-1}, OH bending near 1373 cm^{-1}, ether bond vibration near 1150 cm^{-1} (weak), and strong absorptions associated with the saccharide ring modes between 1000 and 1100 cm^{-1}. The heparin, the chondroitin sulfate, and the proteoglycan mimetic graft copolymers should have strong sulfate-stretching absorptions between 1200 and 1250 cm^{-1}. The relative strength of these vibrations can give some indication of the degree of substitution or graft density of the graft copolymers. However, it is important to have very dry samples, as these absorptions may also be sensitive to the degree of hydration [26].
3. Proton nuclear magnetic resonance (^1H NMR) can also be used to confirm the chemistry of the neat polymers and the intermediates (*see* **Note 24**). Dissolve samples in D_2O at concentrations of 1–5 mg/mL depending on sample solubility (*see* **Note 25**). Collect spectra using >64 scans, 5-s relaxation time, at 25 °C or at elevated temperature to improve peak resolution (*see* **Note 26**). The characteristic peaks that should be observed at various stages of the reaction are (D_2O, 400 MHz):

Unmodified HA has a [1H] NMR spectrum similar to that reported by Pomin [27]: δ 4.70–4.30 (m, *H*C1), 3.95–3.25 (m, *H*C2-6), 2.02 (s, -C(O)-CH_3). HA-SH intermediate [28]: [1H] NMR (D_{2O}, 400 MHz): δ 4.65–4.35 (m, *H*C1), 3.95–3.25 (m, *H*C2-6), 3.25–3.15 (m, -CH_2-CH_2-SH), 2.94–2.88 (m, -CH_2-SH), 2.03 (s, -C(O)-CH_3). HA-BMPH intermediate: δ 3.95–3.15 (m, *H*C2-6 and -CH_2-CH_2-S-), 3.25–3.15, 2.9–2.85 (m, -CH_2S-), 2.48–2.40 (m, N-CH_2-), 2.18–2.09 (m, -CH_2-C(O)-hydrazide) 2.03 (s, -C(O)-CH_3).

4. Use dynamic light scattering (DLS) to measure the effective hydrodynamic diameter of the product, and use electrophoretic light scattering (ELS) to measure its zeta potential.
5. For DLS, dissolve samples in PBS at 5 mg/mL (*see* **Note 27**). Pipet the sample into a cuvette. Place the cuvette in the instrument, and allow the temperature to equilibrate. Measure at 25 °C with a fixed angle of 90° (*see* **Note 28**). Increasing degree of substitution or graft density should cause an increase in the hydrodynamic diameter to several hundred nanometers for the most highly grafted copolymers.
6. The same samples used for DLS can generally be used for zeta potential measurements. Pipet the sample into a zeta potential cuvette or insert the electrode. Place the cuvette in the instrument, and collect data at 25 °C (*see* **Note 28**). In general, zeta potential should decrease (become more negative) with increasing graft density.

3.5 Conclusions

A protocol for synthesizing and characterizing PG mimetic graft copolymers, which allows the graft density to be tuned, has been outlined here. This method is versatile and can easily be adapted to a variety of carbohydrate polymers or to peptides. The product synthesized can be tuned depending on polymer size and by controlling grafting density.

4 Notes

1. Hyaluronan is ideal for use as the backbone due to readily modified carboxylic acid substituents along the polymer chain and its extended conformation in solution. However, other polymers could be used. Polyacrylic acid also contains carboxylic acid substituents and dextran can be modified to present carboxylic acid groups using the methods found in Damodaran et al. [28]. When choosing a backbone be sure to use either a polyanionic or neutral polymer.

2. Molecular weight of the polymer backbone will affect the overall size of the PG mimic and may affect the coupling efficiency. A 740 kDa HA is used here successfully.
3. BMPH has a thiol-reactive maleimide group on one end and a hydrazide on the other end. The hydrazide group is reactive with aldehydes and has a lower p*K*a (4–5) compared to primary amines (10–11) [29]. The low p*K*a allows this group to remain uncharged at neutral pH whereas a primary amine would carry a positive charge causing complexation with the GAG side chains. BMPH has a short linker (8.1 Å); other coupling agents that have these functional groups, but different linker lengths, may be available.
4. The 10 mL spin column was the largest produced at the time this document was written. It is large enough for 5 mL of sample. Ten of these are used to remove TCEP and exchange the buffer in the scenario presented here. If a different batch size is used fewer columns or a different column size may be appropriate.
5. CS and heparin are used as the side chains here because they are commonly found in large proteoglycans. Theoretically any carbohydrate could be used. It is expected that a negatively charged polymer that takes on an extended confirmation in solution will have superior coupling efficiency.
6. 300 kDa dialysis tubing is recommended to remove any uncoupled CS and heparin (84 kDa and 14 kDa respectively) from the HA backbone (740 kDa). If a side chain or backbone with different molecular weights are used, a different pore size may be required.
7. For all steps that instruct the user to dissolve, place a stir bar inside the flask or beaker and position in the center of a stir plate. Turn on the stirring mechanism and slowly increase the speed until a funnel forms (usually 200–400 rpm). Use a stir bar that is small enough to lay flat and spin without constraint in the vessel.
8. 250 mg of HA was chosen for the purposes described in this book chapter. This provides enough material to make several different PG mimics. Different masses may be required for different purposes. Scale the reagents accordingly.
9. Stoppers keep the solvent from evaporating. Place a needle (18 gauge works fine) into the septum of stopper to relieve the pressure while heating.
10. Remove the stopper to add dry reagents.
11. It is easiest to raise the pH by first dissolving sodium hydroxide in diH_2O at 0.5 M and then adding this solution drop-wise until the desired pH is reached.

12. Use at least a 100:1 buffer-to-sample volume ratio to maintain the concentration gradient. Measure plenty of tubing, a few inches longer than the beaker is tall. Soak the dialysis tubing in buffer until it is wet and pliable. The time necessary varies depending upon brand of tubing. Fold over the bottom end of the tubing three times and seal with a clip. Tying the end in a knot will work, but there is a risk of tearing the tubing.
13. Dialysis is driven by a concentration gradient. Water will enter the tubing during this process. Leave enough space in the dialysis tubing for the volume to expand. Generally, filling halfway is sufficient. To speed up the process, a lower beginning salt concentration may be used and buffer may be changed more frequently, but this will require more space in the dialysis tubing. Sometimes it is necessary to empty the dialysis tubing and split the contents into two (or more) fresh sets of tubing to provide enough space.
14. The solution can either be frozen overnight at −20 °C or quickly using liquid nitrogen. The different temperatures and freezing speeds lead to different textures in solid product due to different crystal sizes formed during freezing. This does not affect the quality of the final product. Be sure that the solution is completely frozen and a strong vacuum is being pulled without leaks or lyophilization will not work.
15. The thiol content on the HA backbone can be determined using the Ellman's reagent test. First prepare the reaction buffer, 0.1 M sodium phosphate, pH 8, with 1 mM EDTA. EDTA dissolves better at high pH and will lower the pH of the solution as it dissolves. Then prepare standards by dissolving cysteine hydrochloride monohydrate in the reaction buffer in concentrations ranging from 0 to 1.5 mM (0, 0.25, 0.5 0.75, 1.0, 1.25, and 1.5 mM). Ellman's reagent test should be performed following reduction with TCEP described in Subheading 3.2 using the reaction buffer as the exchange buffer in the desalting column. An initial estimate of 50 % thiolation is appropriate for calculating the necessary amount of TCEP required for reduction prior to Ellman's reagent test:

    Dissolve 4 mg of Ellman's reagent in 1 ml of reaction buffer to make stock solution.

    Set up enough tubes for each standard and your sample; add 2.5 ml of reaction buffer and 50 μl of Ellman's reagent stock solution to each tube.

    Calculate your expected concentration of thiol from your sample, assuming 50 % thiolation of the HA backbone. If this concentration is greater than 1.5 mM, then dilute your sample with reaction buffer so that the expected concentration, assuming 50 % thiolation, is 1 mM. This should put your sample within the detection range of the assay. It is wise to test several concentrations of your sample.

Add 250 μl of standard or sample to each tube.

Incubate at room temperature for 15 min.

Pipette 100 μl of each into a 96-well plate in triplicate and read absorbance in a plate reader at 412 nm.

Plot the results for the standards against their concentrations to create a standard curve. Use this standard curve to determine the concentration of the sample and degree of thiolation.

This degree of thiolation is used throughout this protocol to determine appropriate amounts of reagents to achieve desired grafting densities.

16. Always balance the centrifuge.
17. The tops of the spin columns often bend under the force exerted by centrifuging causing the caps to fracture. This does not damage the resin bed or the sample. It is acceptable to perform centrifugation without the cap.
18. These masses are for one side chain to every thiol calculated from the Ellman's reagent test. It is assumed that the BMPH coupling efficiency is 100 % and that the number of thiols is equivalent to the number of available hydrazide groups. The mass of GAG side chain added will change depending upon the GAG molecular weight and the desired grafting density. Grafting density and size of PG mimic can be tuned by changing the ratio of side chains to available hydrazides.
19. The reducing agent, STAB, is sensitive to water. Store STAB in a desiccator or under nitrogen. Because of this sensitivity, glassware should by dry, solvents should be anhydrous, and the reaction should be sealed and purged with nitrogen. Thirty minutes of purging is sufficient. When transferring solvent do not remove the lid; use a syringe to draw and expel the solvent through the septum.
20. The purpose of the solvent trap is to collect harmful solvents to protect the pump and the user. DMF has a high boiling point, 153 °C at atmospheric pressure (~60 °C under vacuum). Either liquid nitrogen or dry ice and isopropanol will be cold enough to condense it. Check the level of the bath and refill it at least every 24 h. This time point will depend upon the insulation of the Dewar flask used (a porous foam lid improves insulation). Ensure that the vessel is sealed; if air is being drawn in liquid oxygen can form.
21. The product will appear yellow and sticky. It does not need to be completely dissolved to move on to the next step. Use enough water to get the product off of the sides of the flask.

Several rinses may be necessary. If large volumes of water are used there will be a large volume to dialyze and lyophilize, which will be time consuming and may result in low recovery.

22. Perform these analyses on raw polymers, intermediates, and the final product to confirm the chemistry through all of the steps.
23. Other forms of FTIR, such as a KBr pellet, are also acceptable. ATR was chosen here because it is nondestructive.
24. ^{13}C NMR could provide additional information.
25. Higher concentrations will result in better signal, but HA dissolves best at 1–2 mg/ml whereas heparin will readily dissolve at 5 mg/ml.
26. More scans and a longer relaxation time will result in better signal-to-noise ratio.
27. Different samples may have different solubility. For light scattering, the solution should be slightly translucent. If the sample is too concentrated, the polymer will aggregate and give false readings. If the solution is too dilute the instrument will not be able to acquire a signal. It is advantageous to start with a more concentrated sample and slowly dilute.
28. The ideal number of counts will vary with the instrument. Check the manual for specifications. Multiple readings are always beneficial. Three is usually fine for DLS and five for zeta potential.

Acknowledgements

Sean M. Kelly contributed to the successful demonstration of this technique. We thank Prof. Patrick A. Johnson (University of Wyoming) for access to dynamic light scattering and electrophoretic mobility instrumentation, Prof. Travis S. Bailey (Colorado State University) for helpful discussions, and Prof. Melissa M. Reynolds and Alec Lutzke for assistance with ATR-FTIR. Funding for the original work done to develop this protocol was provided by the National Science Foundation (DMR 0847641).

References

1. Schaefer L, Schaefer RM (2010) Proteoglycans: from structural compounds to signaling molecules. Cell Tissue Res 339(1):237–246
2. Wight TN (2002) Versican: a versatile extracellular matrix proteoglycan in cell biology. Curr Opin Cell Biol 14(5):617–623
3. Dudhia J (2005) Aggrecan, aging and assembly in articular cartilage. Cell Mol Life Sci 62(19-20):2241–2256
4. Hardingham TE, Fosang AJ (1992) Proteoglycans - many forms and many functions. FASEB J 6(3):861–870

5. Kolset S, Tveit H (2008) Serglycin–structure and biology. Cell Mol Life Sci 65(7–8):1073–1085
6. Gandhi NS, Mancera RL (2008) The structure of glycosaminoglycans and their interactions with proteins. Chem Biol Drug Des 72(6):455–482
7. Boddohi S, Kipper MJ (2010) Engineering nanoassemblies of polysaccharides. Adv Mater 22(28):2998–3016
8. Melrose J, Roughley P, Knox S, Smith S, Lord M, Whitelock J (2006) The structure, location, and function of perlecan, a prominent pericellular proteoglycan of fetal, postnatal, and mature hyaline cartilages. J Biol Chem 281(48):36905–36914
9. Kenagy RD, Plaas AH, Wight TN (2006) Versican degradation and vascular disease. Trends Cardiovasc Med 16(6):209–215
10. Iozzo RV, Murdoch AD (1996) Proteoglycans of the extracellular environment: clues from the gene and protein side offer novel perspectives in molecular diversity and function. FASEB J 10(5):598–614
11. Schönherr E, Järveläinen H, Sandell L, Wight T (1991) Effects of platelet-derived growth factor and transforming growth factor-beta 1 on the synthesis of a large versican-like chondroitin sulfate proteoglycan by arterial smooth muscle cells. J Biol Chem 266(26):17640–17647
12. Weyers A, Linhardt RJ (2013) Neoproteoglycans in tissue engineering. FEBS J 280(10):2511–2522
13. Lee S-G, Brown JM, Rogers CJ, Matson JB, Krishnamurthy C, Rawat M, Hsieh-Wilson LC (2010) End-functionalized glycopolymers as mimetics of chondroitin sulfate proteoglycans. Chem Sci 1(3):322–325
14. Paderi JE, Panitch A (2008) Design of a synthetic collagen-binding peptidoglycan that modulates collagen fibrillogenesis. Biomacromolecules 9(9):2562–2566
15. Paderi JE, Sistiabudi R, Ivanisevic A, Panitch A (2009) Collagen-binding peptidoglycans: a biomimetic approach to modulate collagen fibrillogenesis for tissue engineering applications. Tissue Eng Part A 15(10):2991–2999
16. Kishore V, Paderi JE, Akkus A, Smith KM, Balachandran D, Beaudoin S, Panitch A, Akkus O (2011) Incorporation of a decorin biomimetic enhances the mechanical properties of electrochemically aligned collagen threads. Acta Biomater 7(6):2428–2436
17. Sharma S, Panitch A, Neu CP (2013) Incorporation of an aggrecan mimic prevents proteolytic degradation of anisotropic cartilage analogs. Acta Biomater 9(1):4618–4625
18. Bernhard JC, Panitch A (2012) Synthesis and characterization of an aggrecan mimic. Acta Biomater 8(4):1543–1550
19. Su L, Zhao Y, Chen G, Jiang M (2012) Polymeric vesicles mimicking glycocalyx (PV-Gx) for studying carbohydrate–protein interactions in solution. Polym Chem 3(6):1560–1566
20. Sarkar S, Lightfoot-Vidal SE, Schauer CL, Vresilovic E, Marcolongo M (2012) Terminal-end functionalization of chondroitin sulfate for the synthesis of biomimetic proteoglycans. Carbohydr Polym 90(1):431–440
21. Boddohi S, Almodóvar J, Zhang H, Johnson PA, Kipper MJ (2010) Layer-by-layer assembly of polysaccharide-based nanostructured surfaces containing polyelectrolyte complex nanoparticles. Colloids Surf B 77:60–68
22. Boddohi S, Moore N, Johnson PA, Kipper MJ (2009) Polysaccharide-based polyelectrolyte complex nanoparticles from chitosan, heparin, and hyaluronan. Biomacromolecules 10:1402–1409
23. Place LW, Sekyi M, Kipper MJ (2014) Aggrecan-mimetic, glycosaminoglycan-containing nanoparticles for growth factor stabilization and delivery. Biomacromolecules 15:680–689
24. Volpato FZ, Almodovar J, Erickson K, Popat KC, Migliaresi C, Kipper MJ (2012) Preservation of FGF-2 bioactivity using heparin-based nanoparticles, and their delivery from electrospun chitosan fibers. Acta Biomater 8(4):1551–1559
25. Place LW, Kelly SM, Kipper MJ (2014) Synthesis and characterization of proteoglycan-mimetic graft copolymers with tunable glycosaminoglycan density. Biomacromolecules 15:3772–3780
26. Servaty R, Schiller J, Binder H, Arnold K (2001) Hydration of polymeric components of cartilage—an infrared spectroscopic study on hyaluronic acid and chondroitin sulfate. Int J Biol Macromol 28(2):121–127
27. Pomin VH (2013) NMR chemical shifts in structural biology of glycosaminoglycans. Anal Chem 86(1):65–94
28. Damodaran VB, Place LW, Kipper MJ, Reynolds MM (2012) Enzymatically degradable nitric oxide releasing S-nitrosated dextran thiomers for biomedical applications. J Mater Chem 22(43):23038–23048
29. Raddatz S, Mueller-Ibeler J, Kluge J, Wass L, Burdinski G, Havens JR, Onofrey TJ, Wang D, Schweitzer M (2002) Hydrazide oligonucleotides: new chemical modification for chip array attachment and conjugation. Nucleic Acids Res 30(21):4793–4802

Part II

Glycopolymer Nanoparticle Conjugates

Chapter 8

Galactosylated Polymer Nano-objects by Polymerization-Induced Self-Assembly, Potential Drug Nanocarriers

Mona Semsarilar, Irene Canton, and Vincent Ladmiral

Abstract

Glycopolymer-based nanostructures are invaluable tools to both study biological phenomena and to design future targeted drug delivery systems. Polymerization-induced self-assembly, especially RAFT aqueous dispersion polymerization is a unique method to prepare such polymer nanostructures, as it enables the preparation of very-well-defined morphologies at very high concentrations. Here we describe the implementation of PISA to the synthesis of galactosylated spheres, wormlike micelles and vesicles, and the preliminary results of cell toxicity, cell uptake, and cargo delivering capacity of galactose-decorated vesicles.

Key words Glycopolymers, Polymerization-induced self-assembly, RAFT polymerization, Cell uptake

1 Introduction

Incorporating a biomimetic design on novel nanomaterials is one of the most promising scientific and technological challenges of the coming years, as it is predicted to be beneficial in tackling clinical problems. The benefit of using design principles tried and tested by nature has inspired many polymer chemists to adopt approaches that are biologically inspired. In this way, synthetic polymer drug carriers that mimic biological membrane enclosed structures have been created as reservoir for drug formulations owing to their ability to improve the stability, solubility, and the bioavailability of pharmaceutical molecules [1]. Furthermore, synthetic polymeric drug carriers can be engineered so as to target specific biological loci where the cargo is needed. Glyco-targeting for example offers a new avenue for the design of improved targeting systems. Glyco-targeting takes advantage of the highly specific interactions of particular glycan receptors with their carbohydrate ligands. These interactions could potentially outperform in specificity and affinity many other ligand-binding-based drug delivery systems, given their biochemical complexity [2]. Glycopolymers (synthetic polymers displaying carbohydrate ligands) are thus very

Xue-Long Sun (ed.), *Macro-Glycoligands: Methods and Protocols*, Methods in Molecular Biology, vol. 1367,
DOI 10.1007/978-1-4939-3130-9_8, © Springer Science+Business Media New York 2016

interesting materials. Thanks to progress in polymerization techniques allowing the control over the composition and architectures of synthetic macromolecules, it is now possible to design well-defined amphiphilic block copolymers comprising of a hydrophilic glycopolymer segment and a hydrophobic block. Amphiphilic block copolymers can self-assemble in water into a variety of morphologies depending on the nature of the blocks, and on the volume fraction of each block. In particular, it is possible to fabricate vesicles or polymersomes from block copolymers. Such polymer morphologies are very attractive for drug-delivery strategies compared to small molecules vesicles such as liposomes for their enhanced stability in dilute conditions, and the functionalization possibilities they offer. Recently, polymerization-induced self-assembly was demonstrated to be a very powerful technique to produce pure phase block copolymer morphologies at high solids contents. Here we describe the methods we used to prepare well-defined glycopolymer-containing block copolymers, to self-assemble these macromolecules and to start assessing the in vitro interactions of the self-assembled structures with live cells.

2 Materials

2.1 Chemicals

1. Thio-β-D-galactose (GalSH) is prepared according to the method described by Floyd et al. [3] in an overall yield of 70 %.
2. Galactose methacrylate (GalSMA) (synthesized from galactose and 3-(acryloyloxy)-2-hydroxypropyl methacrylate as described in Subheading 3).
3. 4-Cyano-4-(((phenethylthio)carbonothioyl)thio)pentanoic acid (PETTC) (synthesized in-house as described in the method section).
4. 3-(Acryloyloxy)-2-hydroxypropyl methacrylate (Sigma-Aldrich).
5. Azobis-4-cyanopentanoic acid (ACVA, >98 %) (Fluka).
6. Glycerol monomethacrylate (Hythe, UK).
7. 2-Hydroxypropyl methacrylate (HPMA, 97 %) (Hythe, UK) (*see* **Note 1**).
8. Dimethylphenylphosphine (Sigma-Aldrich).
9. 2-Phenylethanethiol (Sigma-Aldrich).
10. Sodium hydride 60 % in oil (Sigma-Aldrich).
11. Carbon disulfide (Sigma-Aldrich).
12. Solid iodine (Sigma-Aldrich).
13. Sodium thiosulfate (Sigma-Aldrich).

14. DMEM medium (Biosera UK).
15. Foetal calf serum (Biosera UK).
16. L-Glutamine 200 mM solution (Sigma-Aldrich).
17. Penicillin-Streptomycin solution (100×) (Sigma-Aldrich).
18. Amphotericin B solution (Sigma-Aldrich).
19. Trypsin-EDTA.
20. Rhodamine B octadecyl ester perchlorate (Sigma-Aldrich).
21. 3-(4,5-Dimethyl-2-thiazolyl)-2,5-diphenyl-2H-tetrazolium bromide (MTT).
22. Hoechst 33342 solution (Thermo Scientific, UK).

2.2 Other Commercial Reagents and Solvents

1. Silica gel 60 (0.0632–0.2 mm) (Darmstadt, Germany).
2. Dialysis membrane (molecular weight cutoff, MWCO = 1000) (Fisher Scientific, UK).
3. D_2O (Goss Scientific Instruments Ltd., UK).
4. CD_3OD (Goss Scientific Instruments Ltd., UK).
5. Uranyl formate (Polysciences).
6. Ricinus communis (castor bean) Agglutinin RCA_{120} (Sigma).

2.3 Buffers and Coating Solutions

1. 150 mM Phosphate buffer, pH 7.2.
2. HEPES buffer: 10 mM HEPES, 150 mM NaCl, 1 mM $MnCl_2$, 1 mM $CaCl_2$, pH 7.4.
3. PBS (Oxoid, UK).

2.4 Cell Line

1. Primary human dermal fibroblasts (HDFs) (LGC standards, Teddington, UK): Cells are maintained in DMEM (Biosera, UK) supplemented with 10 % v/v fetal calf serum, 2 mM L- glutamine, 100 IU/mL penicillin, 100 mg/mL streptomycin, and 0.625 μg/mL amphotericin B (all from Sigma-Aldrich, UK).
2. Cells are replenished with fresh medium twice a week and subcultured routinely using 0.02 % (w/v) trypsin-EDTA (Sigma-Aldrich, UK).
3. Cells are used for experimentation between passages 4 and 8.

2.5 Gel Permeation Chromatography

Homopolymer and diblock copolymer molecular weight distributions are determined by DMF GPC.

1. The gel permeation chromatography (GPC) setup comprises two polymer laboratories PL gel 5 μm Mixed-C columns maintained at 60 °C in series with a Varian 390 LC refractive index detector.

2. The flow rate is 1.0 mL/min, and the mobile phase contained 10 mM LiBr.
3. Ten near-monodisperse PMMA standards (M_p = 625 to 618,000 g/mol) are used for calibration.

2.6 ^{1}H NMR Spectroscopy

1. All ^{1}H NMR and ^{13}C NMR spectra are recorded in CD_3OD, d_6-DMSO, or D_2O.
2. A 250 MHz Bruker Avance 250 or a 400 MHz Bruker Avance 400 spectrometer is used.

2.7 Transmission Electron Microscopy

1. TEM images were acquired using a Philips CM100 instrument operating under UHV at 100 kV.
2. To prepare TEM samples, 5.0 μL of a dilute aqueous copolymer solution was placed onto a carbon-coated copper grid, stained using uranyl formate solution, and then dried under ambient conditions.

2.8 Dynamic Light Scattering

1. DLS measurements were conducted at 25 °C using a scattering angle of 173° with a Malvern Instruments Zetasizer Nanoseries instrument equipped with a 4 mW He-Ne laser operating at 633 nm, an avalanche photodiode detector with high quantum efficiency, and an ALV/LSE-5003 multiple tau digital correlator electronics system.
2. The intensity-average diameter and polydispersity of the diblock copolymer particles were calculated by cumulants analysis of the experimental correlation function using Dispersion Technology Software version 6.20.

2.9 UV-Visible Spectroscopy

1. Turbidimetry studies were conducted at 20 °C using a Cary 50 UV-visible spectrophotometer at a wavelength of 420 nm.
2. All lectin interaction studies were performed in HEPES buffer (HEPES 10 mM, NaCl 150 mM, $MnCl_2$ 1 mM, $CaCl_2$ 1 mM) at pH 7.4.
3. Negative control: A cuvette containing 0.50 mL of a 2 μM RCA120 solution in HEPES buffer was placed in the spectrometer. 0.50 mL aliquots of 50 μM homopolymer solutions (either $PGMA_{51}$ or $PGalSMA_{34}$) were added to the cuvette and the absorbance at 420 nm was monitored over time.
4. Lectin assay: A cuvette containing 0.50 mL of a 2 μM solution of RCA_{120} in HEPES buffer was placed in the spectrometer. 0.50 mL of 1.0 wt% aqueous diblock copolymer dispersion was added to the cuvette and the absorbance at 420 nm was monitored over time.

3 Methods

The galactose methacrylate monomer GalSMA (Fig. 1) was synthesized from 1-thio-β-D-galactose and 3-(acryloyloxy)-2-hydroxypropyl methacrylate using Michael addition catalyzed by dimethylphenylphosphine (Fig. 1). This very efficient reaction allows the preparation of multigram quantities of saccharide-functionalized methacrylate from a relatively easy-to-synthesize thiol. The method is very general and should allow the synthesis of functional methacrylate monomers from any thiols with high yields.

The RAFT agent, 4-cyano-4-(2-phenylethane sulfanylthiocarbonyl) sulfanyl pentanoic acid (PETTC), was designed to polymerize methacrylate monomers with good control and to be easy to synthesize. As such, a trithiocarbonate featuring the 4-cyanovaleric acid R group was chosen. The phenylethyl thiol Z-group provides very convenient ^{1}H NMR signals for end-group analysis and molecular weight calculations. The synthesis is very straightforward (Fig. 2), and PETTC can be isolated without recourse to column chromatography.

RAFT polymerizations of GalSMA and GMA were carried out in methanol and ethanol, respectively, under commonly used reaction conditions. The polymerizations were stopped at high conversions (ca. 90 %) and the polymers were easily purified by dialysis and isolated as pure yellow solids after lyophilization.

Polymerization-induced self-assembly was achieved by RAFT polymerization under aqueous dispersion conditions of 2-hydroxypropyl methacrylate (HPMA) using PGalSMA and PGMA as macroRAFT agents. This polymerization protocol is a very efficient way to synthesize well-defined amphiphilic block copolymer ($M_n/M_w < 1.27$) self-assembled morphologies at high concentrations. HPMA is a very interesting monomer. It is soluble in water up to about 13 % w/w at 20 °C, but sufficiently long PHPMA (DP > 20) is completely water insoluble in these conditions. Phase diagrams or synthesis road maps can be advantageously constructed by systematically varying the total solids contents of the formulations and the DP of the PHPMA block. These diagrams can be used to predict the morphologies. When PGalSMA

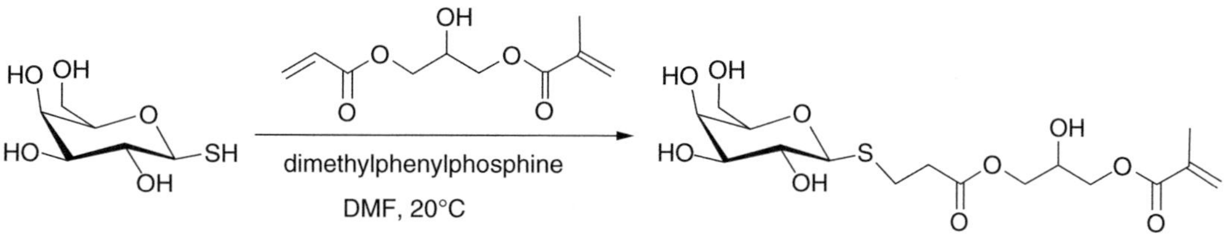

Fig. 1 Synthesis of GalSMA. Reprinted from ref. 4

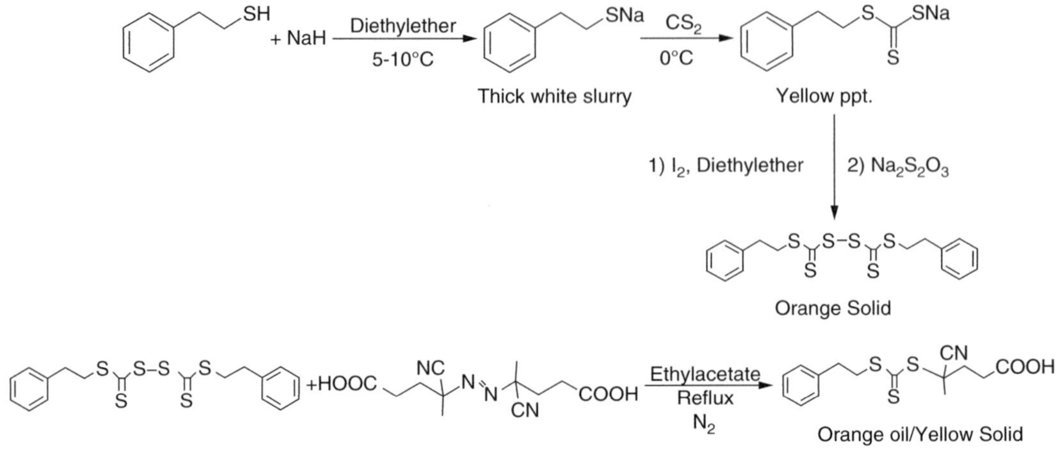

Fig. 2 Synthesis of PETTC

was used as sole macro-RAFT agent the only pure morphologies that could be prepared were spherical particles. Wormlike micelles, thick wall vesicles and tubelike morphologies were also observed but could only be obtained as mixed phases (Fig. 3). On the contrary the use of binary mixtures of macroRAFT agents, such as 1:9 $PGalSMA_{34}+PGMA_{51}$ allows the preparation of well-defined spheres, wormlike micelles and vesicles in pure phases (Fig. 4). This latter diagram was used to prepare galactosylated block copolymer well-defined nanostructures. The bioavailability of the galactose moieties on the surface on these nanostructures was assessed by turbidimetry and dynamic light scattering using RCA_{120}, a galactose-specific lectin (Fig. 5). The specific interaction between RCA_{120} and the galactosylated morphologies provokes the formation of large aggregates which can be easily detected. Finally the phase diagram (Fig. 4) was used to prepare (1:9 $PGalSMA_{34}+PGMA_{51}$)-$PHPMA_{270}$ block copolymer vesicles for cell studies. The RAFT aqueous dispersion polymerization leading to the well-defined vesicles was not carried out in sterile conditions, and the reaction medium, even though the polymerization reaches very high conversion (>98 %) may contain residual chemicals that could be toxic to living cells. The block copolymer vesicles suspension thus dialyzed and lyophilisated to obtain pure block copolymers from which vesicles could be reformed under sterile conditions. The biocompatibility, cellular uptake, and cargo delivery capabilities of these reformed vesicles were then examined using HDF cells. HDF Cells constitute a very useful and sensitive model to assess toxicity induced by nanoparticles [5] and are known to express galectins which avidly bind β-galactosides [6]. MTT-ESTA assay was thus performed and confirmed cell viabilities superior to 95 % over a large vesicle concentration range (from 0.1 to 1 mg/mL) (Fig. 6).

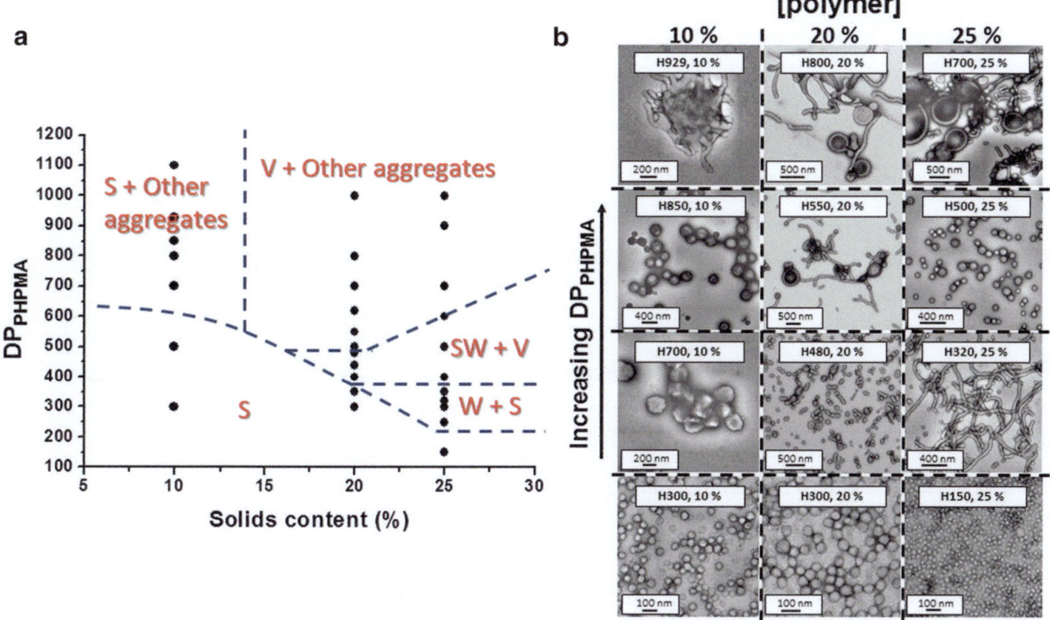

Fig. 3 (**a**) Phase diagram constructed for PGalSMA$_{34}$-PHPMA*x* diblock copolymer nano-objects prepared by RAFT aqueous dispersion polymerization at 70 °C. The target PHPMA DP and the total solids content were systematically varied and the postmortem copolymer morphologies obtained at >98 % HPMA conversion were determined by TEM. N.B. S, SW, W, V, and FS denote spheres, short wormlike micelles, worm-like micelles, vesicles, and frustrated (i.e., kinetically trapped) spheres, respectively. (**b**) Representative TEM images obtained for (1:9 PGalSMA$_{34}$ + PGMA$_{51}$)-PHPMA*x* copolymer nano-objects prepared by RAFT aqueous dispersion polymerization of HPMA at 70 °C. The targeted DP (x) for the PHPMA block (herein denoted by 'H' for brevity) and the copolymer solids content % is indicated on each image. Reprinted from ref. 4

The cellular uptake and drug carrier capability of the highly biocompatible (1:9 PGalSMA$_{34}$ + PGMA$_{51}$)-PHPMA$_{270}$ vesicles were tested using rhodamine B octadecyl ester as a cargo. As seen in Fig. 7, these dye-loaded vesicles were avidly internalized by HDF cells, and the release of the rhodamine dye provoked extensive staining of the cell membranes. The free (in the absence of a carrier) amphiphilic rhodamine dye cannot enter HDF cells [7]. The cellular uptake of the vesicles and the intracellular delivery of their cargo were thus confirmed by the staining of the intracellular compartment. Surprisingly, the staining of the endomembrane system, including the nuclear membrane (Fig. 7) and even co-staining within the nuclear region, was observed. The nuclear region lacks endolysosomal compartments. The staining of this region thus suggests the release of the dye inside the cell and its escape from the normal endocytic pathway.

Galectin-mediated receptor turnover is known to be a rapid process that can potentially avoid degradation in the lysosomal environment [8, 9]. Nevertheless, the apparent release of the rhodamine dye from the vesicles and its location outside the endocytic

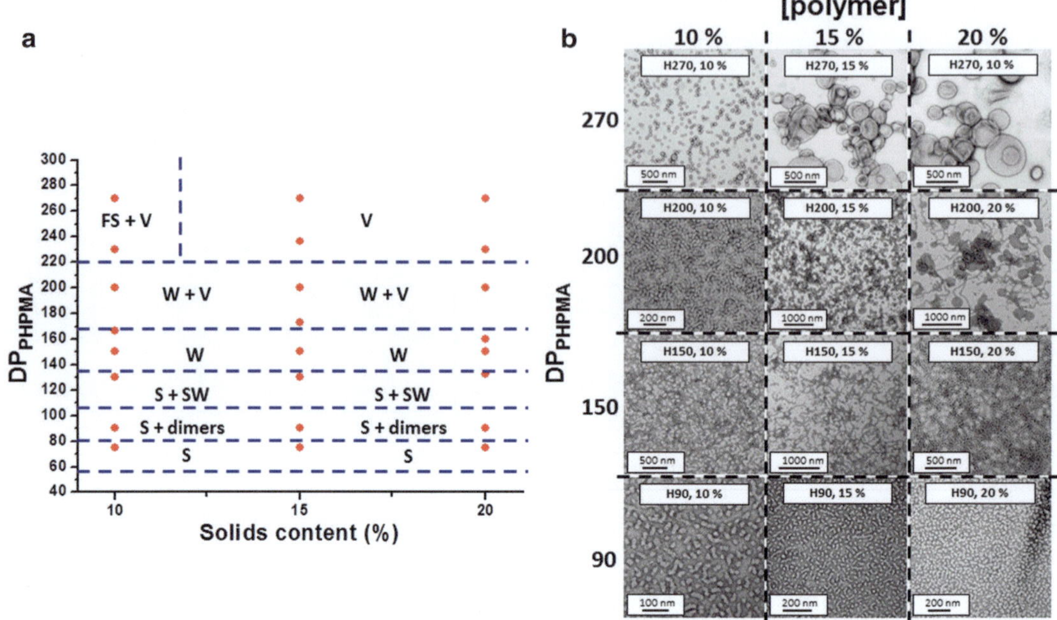

Fig. 4 (**a**) Phase diagram constructed for (1:9 $PGalSMA_{34}$ + $PGMA_{51}$)-$PHPMA_x$ diblock copolymer nano-objects prepared by RAFT aqueous dispersion polymerization at 70 °C. The target PHPMA DP and the total solids content were systematically varied and the post mortem copolymer morphologies obtained at >98 % HPMA conversion were determined by TEM. N.B. S, SW, W, V, and FS denote spheres, short worm-like micelles, worm-like micelles, vesicles, and frustrated (i.e., kinetically trapped) spheres, respectively. (**b**) Representative TEM images obtained for (1:9 $PGalSMA_{34}$ + $PGMA_{51}$)-$PHPMA_x$ copolymer nano-objects prepared by RAFT aqueous dispersion polymerization of HPMA at 70 °C. The targeted DP (x) for the PHPMA block (herein denoted by 'H' for brevity) and the copolymer solids content % is indicated on each image. Reprinted from ref. 4

compartments (within the cell nuclei) is rather surprising. Previous studies suggest that the stability of self-assembled structures based on diblock copolymers containing weakly hydrophobic PHPMA chains [10] is concentration dependent. We thus suggest that in our case, the vesicle dissociation and the release of the rhodamine dye inside the cells are triggered by dilution. Interestingly, other biocompatible methacrylic diblock copolymers based on significantly more hydrophobic chains than PHPMA do not necessarily exhibit this release mechanism [11, 12].

Procedures are to be carried out at room temperature unless otherwise stated. Sterile procedures are to be carried out in a class II laminar flow sterile cabinet.

3.1 Synthesis of Galactose Methacrylate (GalSMA)

1. Place GalSH (5.00 g, 25.48 mmol) in a round-bottom flask and dissolve in DMF (15 mL).
2. Add a solution of 3-(acryloyloxy)-2-hydroxypropyl methacrylate (6.00 g, 28.03 mmol) in DMF (5.0 mL) to the GalSH solution.

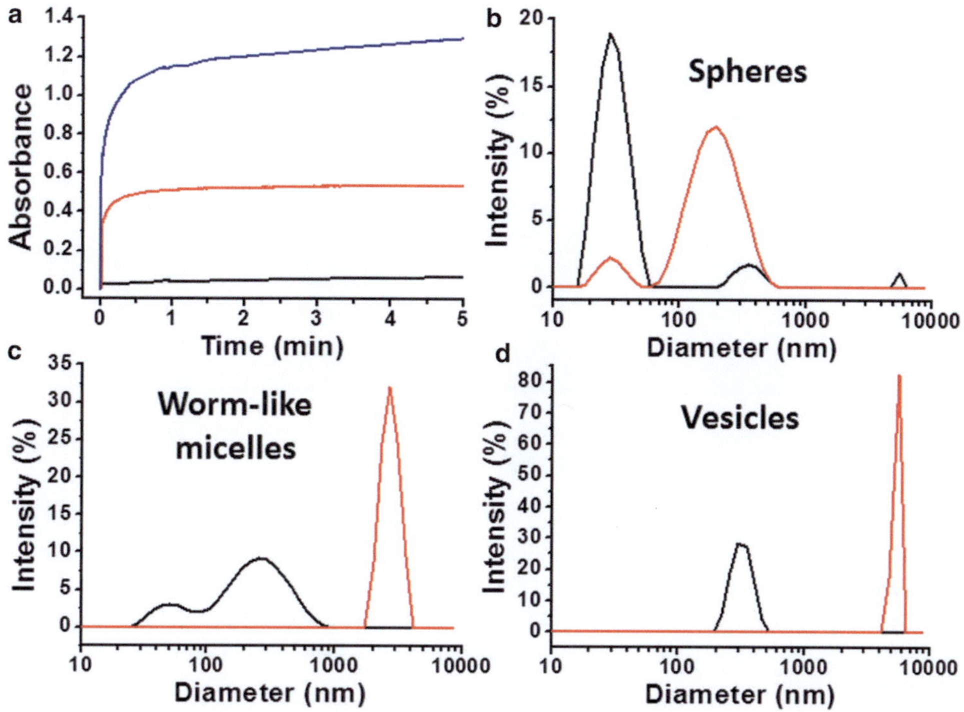

Fig. 5 Galactose-specific lectin interactions with three types of galactose-functionalized diblock copolymer nano-objects (each originally prepared at 20 % solids). (**a**) Turbidimetric assays for (1:9 $PGalSMA_{34}$ + $PGMA_{51}$)-$PHPMA_{75}$ spheres (*black curve*), (1:9 $PGalSMA_{34}$ + $PGMA_{51}$)-$PHPMA_{150}$ wormlike micelles (*red curve*), and (1:9 $PGalSMA_{34}$ + $PGMA_{51}$)-$PHPMA_{270}$ vesicles (*blue curve*). DLS size distributions recorded for the same nano-objects: (**b**) spheres, (**c**) wormlike micelles, and (**d**) vesicles, recorded both in the absence (*black curves*) and presence (*red curves*) of RCA_{120}. Assay conditions: [RCA_{120}] = 1 μM and [copolymer] = 0.50 wt% in 10 mM HEPES buffer at pH 7.2. Reprinted from ref. 4

3. Add dimethylphenylphosphine (10 μL, 7.0×10^{-2} mmol) to the reaction solution.
4. Add the DMF solution dropwise and under vigorous stirring into a large excess (200 mL) of diethyl ether. This results in a white precipitate.
5. Isolate the white precipitate by filtration using a sintered glass funnel (porosity 3).
6. Redissolve this hygroscopic (which may become wet and thus appear as a paste) solid in DMF (15 mL) and precipitate again into diethyl ether as in **step 4**.
7. Filter the precipitate as described in **step 5**, and wash it with diethyl ether (100 mL).
8. Dry the resulting white solid at 20 °C under vacuum in a vacuum oven.
9. Purify the resulting viscous oil by flash chromatography using 9:1 methanol/dichloromethane (*see* **Note 2**).

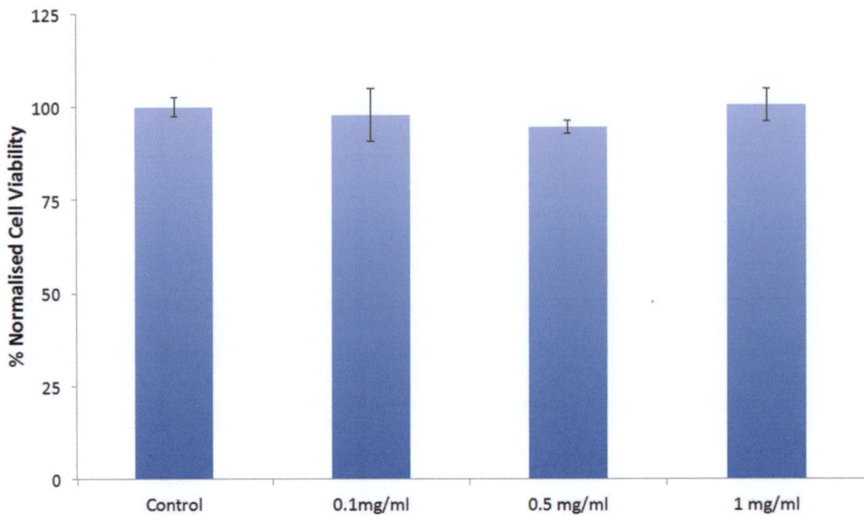

Fig. 6 Normalized cellular viability of human dermal fibroblasts after incubation with (1:9 $PGalSMA_{34}$ + $PGMA_{51}$)-$PHPMA_{270}$ polymersomes. Cells were incubated in the presence of increasing concentrations of polymersomes in cell media over 24 h. Cell viabilities were evaluated using an MTT-ESTA assay and the data were normalized relative to the untreated control (100 % viability). $N = 3$ independent experiments were performed in triplicate wells. Reprinted from ref. 4

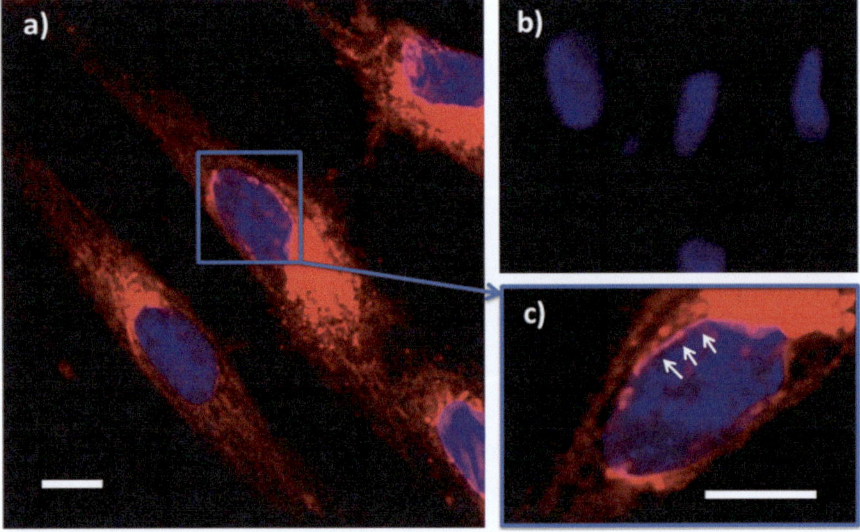

Fig. 7 Effective intracellular delivery of rhodamine B octadecyl ester in human dermal fibroblast (HDF) cells mediated by (1:9 $PGalSMA_{34}$ + $PGMA_{51}$)-$PHPMA_{270}$ vesicles. Cells were incubated for 16 h with 1.0 mg/mL rhodamine B octadecyl ester-loaded vesicles. (**a**) Confocal microscopy image of live HDF cells: note the intracellular staining of membranes (*red*) after exposure to the rhodamine-loaded vesicles, cell nuclei are counterstained blue using Hoechst 33342. (**b**) HDF cells treated with the same vesicles containing no rhodamine dye (negative control). (**c**) Higher magnification image obtained for (**a**): effective intracellular delivery of rhodamine dye allows selective staining of the nuclear membrane (*white arrows*). Scale bar: 10 μm. Reprinted from ref. 4

10. After purification, the yield is estimated by ^1H NMR to be around 90 %. The overall yield based on β-D-galactose penta-acetate is 63 %.
11. ^1H NMR (400.13 MHz, D_2O, 298 K) δ (ppm): 1.93 (s, 3H, -CH_3); 2.83 (t, 2H, -CH_2-COO); 2.92–3.06 (m, 2H, -CH2-S); 3.54 (t, 1H, H_2); 3.61–3.64 (dd, 1H, H_3); 3.67–3.82 (m, 4H, H_5, H_6, -CH_2-CHOH-CH_2-); 3.96 (d, 1H, H_4); 4.20–4.30 (m, 4H, -CH_2-CHOH-CH_2-); 4.48 (d, 1H, H_1); 5.73 (s, 1H, vinyl), 6.16 (s, 1H, vinyl).
12. ^{13}C NMR (400.13 MHz, D_2O, 298 K) δ (ppm): 18.0 (CH_3-); 25.8 (-S-CH_2); 35.5 (-S-CH_2-CH_2-); 61.6 (C_6); 65.9 (2C, -CH_2-CHOH-CH_2-); 67.5, 69.4, 70.2, 74.5, 79.5, 86.7 (6C, C_1, C_2, C_3, C_4, C_5, -CH_2-CHOH-CH_2-); 127.8, 136.2 (2C, vinyl), 169.9, 174.7 (2C, carbonyls).
13. ($M+H^+$): calculated mass = 411.1307, actual mass found = 411.1325.

3.2 Synthesis of 4-Cyano-4-(2-phenylethane sulfanylthiocarbonyl) sulfanyl pentanoic acid [13]

1. Add 2-phenylethanethiol (10.5 g, 76.0 mmol) over 10 min to a stirred suspension of sodium hydride (60 % in oil) (3.15 g, 79.0 mmol) in diethyl ether (150 mL) at a temperature between 5 and 10 °C.
2. A vigorous evolution of hydrogen is observed and the grayish suspension turned to thick white slurry of sodium phenylethanethiolate over 30 min.
3. Cool the reaction mixture to 0 °C and add carbon disulfide (6.00 g, 79.0 mmol) portion-wise to provide a thick yellow precipitate of sodium 2-phenylethanetrithiocarbonate which is collected by filtration after 30 min and used in the next step without purification.
4. Add diethyl ether (100 mL) to the obtained sodium 2-phenylethanetrithiocarbonate (11.6 g, 4.09 mmol), and treat this suspension by portion-wise addition of solid iodine (6.30 g, 25.0 mmol).
5. Stir the reaction mixture at room temperature for 1 h and the remove the white sodium iodide by filtration.
6. Wash the yellow-brown filtrate with an aqueous solution of sodium thiosulfate to remove excess iodine, and dry over sodium sulfate and evaporate to leave a residue of bis-(2-phenylethane sulfanylthiocarbonyl) disulfide (~100 % yield).
7. Degas a solution of 4,4′-azobis(4-cyanopentanoic acid) (ACVA) (2.10 g, 75.0 mmol) and bis-(2-phenylethane sulfanylthiocarbonyl) disulfide (2.13 g, 5.0 mmol) in ethyl acetate (50 mL) by nitrogen bubbling and heat it at reflux under N_2 atmosphere for 18 h.

8. After removal of the volatiles under vacuum, wash the crude product with water (5 × 100 mL). Collect the organic phase and concentrate to afford a yellow residue.
9. Purify the residue by silica column (petroleum ether:ethylacetate; 7:3 gradually increasing to 4:6 (v/v)) to afford 4-cyano-4-(2-phenylethane sulfanylthiocarbonyl) sulfanyl pentanoic acid (PETTC) as a yellow oil (yield = 78 %) (*see* **Note 3**).
10. ^{1}H NMR (400.13 MHz, CD_2Cl_2, 298 K) δ (ppm) = 1.89 (3H, -CH_3), 2.34–2.62 (m, 2H, -CH_2), 2.7 (t, 2H, -CH_2), 3.0 (t, 2H, -CH_2), 3.6 (t, 2H, -CH_2), 7.2–7.4 (m, 5H, aromatic).
11. ^{13}C NMR (400.13 MHz, CD_2Cl_2, 298 K) δ (ppm) = 24.2 (CH_3), 29.6 (CH_2CH_2COOH), 30.1(CH_2Ph), 33.1 (CH_2CH_2COOH), 39.9 (SCH_2CH_2Ph), 45.7 ($SCCH_2$), 118.6 (CN), 127.4, 128.8, 129.2, 144.3 (Ph), 177.4 (C = O), 222.2 (C = S).

3.3 RAFT Homopolymerization of Galactose Methacrylate (GalSMA)

1. Place GalSMA (10.65 g of a 77 wt% methanolic solution; 8.20 g, 19.97 mmol) in a single-neck round-bottom flask containing a magnetic bar, PETTC (0.226 g, 666 μmol), and ACVA (18.60 mg, 66.4 μmol; PETTC/ACVA molar ratio = 10). Close the flask using a rubber septum.
2. 22.15 g of 150 mM phosphate buffer solution (pH 7.2) was added and the final solution is degassed by nitrogen bubbling using a long needle and an escape needle.
3. After 30 min, place the round-bottom flask in a preheated oil bath at 70 °C for 150 min.
4. Quench the reaction by putting the flask into a water bath (at 20 °C) and exposing the reaction mixture to air (by removing the rubber septum).
5. Dialyze the reaction solution (MWCO = 1000) against deionized water, followed by freeze-drying overnight to afford the polymer (97 % conversion). DMF GPC analysis gives $M_n = 16{,}300$ g/mol, $M_w/M_n = 1.13$.
6. End-group analysis via ^{1}H NMR spectroscopy indicates a mean degree of polymerization of 34 ($M_n = 14{,}300$ g/mol), which corresponds to a RAFT CTA efficiency of 85 % for the PETTC.

3.4 RAFT Homopolymerization of Glycerol Monomethacrylate (GMA)

1. Add GMA (7.00 g, 43.70 mmol) to a round-bottom flask containing a magnetic bar, PETTC (269.78 mg, 795.00 μmol), and ACVA (22.27 mg, 79.50 μmol). Close the flask using a rubber septum.
2. Add ethanol (7.00 g) to this solution, and then degas it by nitrogen bubbling using a long needle and an escape needle.
3. After 30 min, place the round-bottom flask in a preheated oil bath at 70 °C for 5 h.

4. Quench the reaction by putting the flask into a water bath (at 20 °C) and exposing the reaction mixture to air (by removing the rubber septum)
5. Dialyze the reaction solution (MWCO = 1,000) against deionized water and freeze-dry overnight to afford the polymer (conversion = 88 %).
6. DMF GPC analysis indicates $M_n = 16{,}200$ g/mol and $M_w/M_n = 1.15$.
7. End-group analysis using ^{1}H NMR gives a mean degree of polymerization of 51 ($M_n = 8500$ g/mol). This indicates a RAFT CTA efficiency of 94 % for the PETTC.

3.5 Polymerization-Induced Self-Assembly

A typical RAFT aqueous dispersion polymerization is performed as follows:

1. Add HPMA (377.0 mg, 2.61 mmol, target DP = 201) and deionized water (1.78 mL) to a sample vial containing a magnetic stir bar. Close the vial using a rubber septum.
2. Add $PGMA_{51}$ macro-CTA (100 mg, 11.7 μmol), $PGalSMA_{34}$ macro-CTA (18.7 mg; 1.31 μmol), and ACVA initiator (200 μL of a 13.0 mM aqueous solution; macro-CTA/initiator molar ratio = 5.0) to the vial.
3. Degas the reaction solution by nitrogen bubbling for 15 min using a long needle and an escape needle, and then place it in a preheated oil bath at 70 °C for 6 h.
4. Quench the reaction by putting the flask into a water bath (at 20 °C) and exposing the reaction mixture to air (by removing the rubber septum) (>99 % conversion, as judged by ^{1}H NMR spectroscopy).
5. Similar polymerizations are conducted targeting alternative PHPMA block lengths, which allow access to spherical, worm-like or vesicular copolymer morphologies (*see* **Note 4**).

3.6 Preparation of Sterile (1/9 $PGalSMA_{34}$ + $PGMA_{51}$)-$PHPMA_{270}$ Vesicles via Film Rehydration

The formation of a thin film is important to obtain appropriate solubilization and vesicle budding via stirring method. Precise polymer: solvent mixture ratios are critical to obtain a thin film coated onto the side of the glassware. However, it is important to use a specific diameter glassware so that the evaporated film is never higher than the volume of buffer used (i.e., PBS) to rehydrate the vesicles. Otherwise macroscopic precipitates of the film will form due to incomplete rehydration and "peeling" of the film. When such precipitates appear in the sample, the formation of vesicles is greatly hindered.

1. Dissolve the freeze-dried copolymers (*see* **Note 5**) in powder form in a 2:1 % v/v methanol/chloroform mixture to form a 1 mM solution.

2. Prepare the fluorescent cargo solution (*see* **Note 6**) dissolving 0.05 mM rhodamine B octadecyl ester perchlorate solution in 2:1 % v/v methanol/chloroform.
3. Mix equal volumes of the rhodamine cargo and copolymer solution in a 10 ml glass vial prior sterilization by filtration.
4. Filter the above mixture through a sterile 0.20 μm Nylon Filter (Millipore) in a sterile laminar flow hood. Replace the glass vial cap with a sterile nylon membrane (0.20 μm pore size) carefully attached around the outer rim of the glass vial with autoclave sealing tape. This ensures sterility of the sample when removing from the laminar flow hood while allowing evaporation of the solvents in the vacuum oven. Evaporate the solvent mixture at 37 °C in a vacuum oven until the formation of a thin dry film is observed (typically 6–8 h). Once evaporated, the sterile nylon membrane should be removed and replaced by the sterile vial cap in a sterile environment (*see* **Note 7**).
5. Rehydrate the polymer film under sterile conditions using phosphate buffer saline (100 mM PBS) at pH 7.4 with constant stirring for 5 days to form a 1.0 % w/w copolymer suspension.
6. Sonicate the vesicle dispersion daily for up to 30 min under controlled temperature (20 °C). This step will allow the breakage of multi-lamellar structures to facilitate the formation of vesicles.
7. Harvest the vesicles in sterile conditions via preparative gel permeation chromatography (GPC), using a sterile glass size exclusion column containing aseptic Sepharose 4B and using sterile PBS at pH 7.4 as an eluent (*see* **Note 8**).
8. Fill the sterile column 2/3 of the length with sepharose 4B. Once the PBS flow has stopped and there is no liquid on the bed, add your sample onto the sepharose carefully without allowing the resin to dry or disturbing the meniscus of the resin.
9. Observe the sample being adsorbed into the top of the resin. Immediately after, fill the bed with sterile PBS and start collecting the fractions (*see* **Note 9**).
10. As the sample runs through the column, the elution time of the fraction with free rhodamine cargo will be retarded due to the smaller molecular weight. This will show as a red fraction in the upper part of the column. Start collecting the fractions. An increased turbidity of the collected fractions will reveal presence of vesicles. Pull the turbid samples together. Annotate the final volume of vesicles (*see* **Note 10**).
11. Check the size distribution of the vesicles using dynamic light scattering and/or TEM.

3.7 MTT Assay

The 3-(4,5-dimethyl-2-thiazolyl)-2,5-diphenyl-2H-tetrazolium bromide (MTT) assay is a very standard biocompatibility test used for a broad range of nanoparticles with potential biological applications. It reports on the cellular metabolic activity after exposure to the nanoparticles. Typical toxicity curves are extrapolated from the data analysis representing the % cellular metabolic activity (normalized to the untreated control wells) over a range of increasing concentrations of the material exposed to the cells.

It is strongly recommended to perform detailed toxicity curves for each cell type used in nanoparticle uptake studies, as different cell types may behave differently. Human dermal fibroblasts are highly sensitive primary cells that are able to take in a broad size range of nanoparticles; thus, they are considered a good model for biocompatibility assays [5]. Further, they are known to express galectins, which makes them ideal candidates to study the biocompatibility of these nanoparticles [6].

The polymer concentration range in the toxicity curves is limited by two parameters:

(a) The solids content range of polymer in the PBS solution that efficiently yields the desired nano-objects. These ranges can easily be determined by examining the phase diagrams.

(b) The maximum tolerated PBS percentage in cell medium. As the nanoparticles are suspended in PBS solution, when added to the cells in culture they will dilute the cell medium. Depending on the cell type, concentrations of PBS above 10 % in the medium may produce toxicity and/or stress. It is strongly recommended to include the relevant PBS-only controls with the same volume of PBS added into the wells whenever it is required to go above 10 %.

1. Seed $3–4 \times 10^4$ HDF cells per well of a standard 24-well plate in a final volume of 1 ml of cell medium.
2. Grow the cells in a humidified CO_2 incubator at 37 °C and in a 95 % air/5 % CO_2 environment until 70 % confluence (typically 48 h) (*see* **Note 11**).
3. Change the medium and add the vesicle dilutions to the corresponding wells.
4. Incubate the cells for further 24 h with varying concentrations of vesicles.
5. Afterwards, remove the treatments, wash the wells with PBS, and then incubate with MTT solution (0.50 mg/ml MTT in PBS, 1.0 mL per well for 24-well plates) for 45 min at 37 °C and in a 95 % air/5 % CO_2 environment. The intracellular dehydrogenase activity reduces MTT to form an insoluble purple formazan salt (*see* **Note 12**).

6. After 45 min, aspirate the MTT solution and solubilize and release the intracellular formazan product by adding acidified isopropanol (0.30 mL per well of 24-well plate or 1 mL/cm^2 cultured tissue) incubating for 10 min.
7. Read the optical density at 570 nm using a plate reading spectrophotometer (with a reference filter at 630 nm) (*see* **Note 13**). For statistical analysis (student's *t*-test), experiments should be performed in triplicate wells with a total of $N=3$ independent experiments.

3.8 Cellular Internalization of (1:9 PGalSMA$_{34}$ + PGMA$_{51}$)-PHPMA$_{270}$ Vesicles and Live Imaging

The live imaging experiments were performed using a Zeiss LSM510 Meta instrument (40× magnification).

1. Seed cells on required well size tissue culture plates at an appropriate density to achieve 50–60 % confluent monolayers within 2 days of culture. For HDFs growing in BD Falcon 96-well imaging plates the seeding density is approximately 5×10^3. Scale up/down as required.
2. Grow the cells in a humidified CO_2 incubator at 37 °C and in a 95 % air/5 % CO_2 environment until 50–60 % confluence (typically 48 h).
3. Change the medium and add the rhodamine octadecylester-vesicle treatment at the right concentration. Typically, 1:10 dilution of a 10 mg/mL polymer solution. Scale up/down as required.
4. Incubate the cells with the vesicles overnight (typically 16 h).
5. Wash the cells three times with sterile PBS and make up a 1.0 μg/mL of Hoechst 33342 solution in sterile PBS. Add 100 μL of the stain solution per well of a 96-well plate (scale up/down as required) to stain the cell nuclei and incubate for 10 min.
6. Wash three times with sterile PBS and add 100 μL of imaging medium (the same cell culture medium without phenol red) for subsequent live imaging experiments using fluorescent microscopy (*see* **Note 14**). Table 1 specifies the relevant Ex/Em for the fluorophores as well as the expected biological targets.

4 Notes

1. This chemical comprises approximately 75 % 2-hydroxypropyl methacrylate and 25 mol % 2-hydroxyisopropyl methacrylate. According to HPLC analysis, this monomer also contained about 0.10 mol % dimethacrylate impurity. These impurities may be responsible for the broadening of the SEC-HPLC

Table 1
Dyes and live cell imaging

Dye	Ex/Em (nm)	Target	Supplier	Concentration
Rhodamine B octadecyl ester	554/575	Cell membranes	Sigma-Aldrich	5 % molar ratio
Hoechst 33342	350/461	Nuclei (DNA)	Thermo Scientific, UK	1 μg/mL

chromatograms observed when the polymerizations are taken to high conversions (above 95 %).

2. This monomer could easily homopolymerize spontaneously during the solvent removal steps (when using rotary evaporation) or upon storage. It was thus not isolated, but instead was stored at −20 °C as a 77 % w/w concentrated solution in methanol. Such a solution is stable for at least one month.
3. Pure PETTC could also be isolated via the two following methods:
   - Precipitation of the crude PETTC solution (in ethylacetate) in ice cold cyclohexane.
   - Recrystallization from ethylacetate: Dissolve the crude PETTC in minimum amount of ethylacetate and leave it at −20 °C (estimated duration: 3 weeks). The recrystallization time can be shortened if the solution is seeded with pure PETTC crystals.
4. Targeting relatively long PHPMA blocks at 10 % w/w solids produced a mixture of spheres and small vesicles. These spheres are believed to be kinetically trapped copolymer morphologies. However, no evidence for a sphere-to-vesicle transformation was observed during the long-term storage of these samples at 20 °C for 5 months.
5. The thin-film rehydration technique is used to optimize cell viability. The purification protocol allows the production of ultraclean freeze dried polymer powder (i.e., no small-molecular-weight compounds or volatiles from the reaction solution after polymerization) and allows for facile sterile production of vesicles. The vesicles produced this way have comparable dimensions by DLS to those obtained by PISA. Nevertheless, PISA synthesis can be reproduced under sterile conditions and following thorough sterile dialysis of the samples the vesicles obtained can be used likewise.
6. To facilitate the detection of vesicle uptake during the cell internalization experiments, the vesicles were loaded with a

fluorescent cargo (rhodamine B octadecyl ester perchlorate) (Sigma-Aldrich) that otherwise cannot enter freely the cells.

7. Once the sterile sample has been thoroughly dehydrated and the original sterile cap is back on, it can be stored for several weeks in a cold dry environment protected from the light. It is important to ensure that all the solvent has been evaporated from the film. Failure to remove solvent will produce precipitates, affect cellular viability and reduce significantly the encapsulation efficiency of the rhodamine octadecyl ester cargo.
8. Most manufacturers provide clear information to whether the glass GPC columns can resist high-pressure sterilization by autoclaving. If this is not possible, it is recommended to use aseptic technique thoroughly cleaning the glassware with sterilizing agents (i.e., typical laboratory-use surfactants) and incubating overnight in 70 % ethanol. Following this, it is recommended to profusely wash the columns with sterile deionized water in a class II laminar flow hood. The resin Sepharose 4B cannot be sterilized with high pressure. The resin beads should be spun (4 °C, 5 min, 1000 × *g*) and incubated in 70 % ethanol overnight at 4 °C. Afterwards, the resin should be washed with at least 10 volumes of PBS before use. This will ensure that no solvent is present. Always follow manufacturer's instructions for best results.
9. It is important to wait for the sample to be fully adsorbed into the resin before loading the PBS buffer. Otherwise, the vesicle sample will be diluted.
10. A good indication that the vesicles have been successfully separated from the free cargo is to recover the same volume of solution as the volume loaded with minimal changes in turbidity while observing a red fraction retained in the column (free cargo). Changes in final volume recovered indicate samples loss or dilution.
11. It is important to grow the cells for a minimum of 2 days in the tissue culture plates before testing the nano-vesicle on the 70 % monolayers and therefore the initial cell density should be carefully considered. This is because the proteolytic activity of the enzymatic compounds used to detach cell monolayers can alter cell surface proteins and induce abnormal interactions between the cells and the nanoparticles.
12. The formation of the blue formazan salt can be visualised under an optical microscope. It is recommended to check the control wells (untreated wells) under the microscope after 45 min to ensure effective formation of the formazan salt before solubilization. If the blue salts cannot be observed inside the cells, the plate can be returned to the incubator for

further 15 min. Always follow the manufacturer's instructions for best results.

13. To maintain linearity, the spectroscopy readings should be between 0.05 and 1. If the values are above 1, the samples can be further diluted using acidified isopropanol. The dilution factor should be taken into account when producing the toxicity curves.
14. For optimal results, it is recommended to perform live cell fluorescence microscopy immediately after the samples are ready. The staining with Hoechst 33342 is extremely useful in providing information about the DNA content of living cells, but at there are reports of toxicity including cell death, moderate mutation, and significant cell cycle perturbations with increasing concentrations and prolonged incubations [14].

Acknowledgments

This work was supported by the EPSRC (EP/G007950/1, EP/I012060/1 and EP/E012949/1), and an ERC Advanced Investigator grant awarded to Professor S. P. Armes (PISA 320372).

References

1. Avila-Olias M, Pegoraro C, Battaglia G, Canton I (2013) Inspired by nature: fundamentals in nanotechnology design to overcome biological barriers. Ther Deliv 4:27–43
2. Bertozzi CR, Kiessling LL (2001) Chemical glycobiology. Science 291:2357–2364
3. Floyd N, Vijayakrishnan B, Koeppe JR, Davis BG (2009) Thiyl glycosylation of olefinic proteins: S-linked glycoconjugate synthesis. Angew Chem Int Ed 48:7798–7802
4. Ladmiral V, Semsarilar M, Canton I, Armes SP (2013) Polymerization-induced self-assembly of galactose-functionalized biocompatible diblock copolymers for intracellular delivery. J Am Chem Soc 135:13574–13581, http://pubs.acs.org/doi/full/10.1021/ja407033x
5. Chang JS, Chang KL, Hwang DF, Kong ZL (2007) In vitro cytotoxicitiy of silica nanoparticles at high concentrations strongly depends on the metabolic activity type of the cell line. Environ Sci Technol 41:2064–2068
6. Akimoto Y, Hirabayashi J, Kasai K-I, Hirano H (1995) Expression of the endogenous 14-kDa beta-galactoside-binding lectin galectin in normal human skin. Cell Tissue Res 280:1–10
7. Massignani M, Canton I, Sun T, Hearnden V, MacNeil S, Blanazs A, Armes SP, Lewis A, Battaglia G (2010) Enhanced fluorescence imaging of live cells by effective cytosolic delivery of probes. PLoS One 5:e10459
8. Delacour D, Koch A, Jacob R (2009) The role of galectins in protein trafficking. Traffic 10:1405–1413
9. Stechly L, Morelle W, Dessein A-F, André S, Grard G, Trinel D, Dejonghe M-J, Leteurtre E, Drobecq H, Trugnan G, Gabius HJ, Huet G (2009) Galectin-4-regulated delivery of glycoproteins to the brush border membrane of enterocyte-like cells. Traffic 10:438–450
10. Madsen J, Armes SP, Bertal K, MacNeil S, Lewis AL (2009) Preparation and aqueous solution properties of thermoresponsive biocompatible AB diblock copolymers. Biomacromolecules 10:1875–1887
11. Canton I, Massignani M, Patikarnmonthon N, Chierico L, Robertson J, Renshaw SA, Warren

NJ, Madsen JP, Armes SP, Lewis AL, Battaglia G (2013) Fully synthetic polymer vesicles for intracellular delivery of antibodies in live cells. FASEB J 27:98–108

12. Lomas H, Canton I, MacNeil S, Du J, Armes SP, Ryan AJ, Lewis AL, Battaglia G (2007) Biomimetic pH sensitive polymersomes for efficient DNA encapsulation and delivery. Adv Mater 19:4238–4243

13. Semsarilar M, Ladmiral V, Blanazs A, Armes SP (2012) Anionic polyelectrolyte-stabilized nanoparticles *via* RAFT aqueous dispersion polymerization. Langmuir 28:914–922

14. Durand RE, Olive PLJ (1982) Cytotoxicity, mutagenicity and DNA damage by Hoechst 33342. J Histochem Cytochem 30:111–116

Chapter 9

Synthetic Approach to Biotinylated Glyco-Functionalized Quantum Dots: A New Fluorescent Probes for Biomedical Applications

Christian K. Adokoh, James Darkwa, and Ravin Narain

Abstract

Technological advances that allow deeper penetration in live tissues, such as the development of confocal and the generation of ever-new fluorophores that facilitate bright labeling of cells and tissue components have made imaging of vertebrate model organisms efficient and highly informative. Recently, high luminescence, single-excitation narrow emission, low photo bleaching properties, and low toxicity of high-quality water-soluble QDs have attracted attention for in vivo labeling/imaging of cells. Herein we describe a synthetic approach to biotinylated glycopolymer functionalized quantum dots, with special emphasis on the development of high-quality water-soluble and bioactive QDs with low toxicity for fluorescent probes in biomedical applications.

Key words Quantum dots, Biotinylated glycopolymers, Biotin, Fluorescent probes

1 Introduction

Cellular labeling with fluorescent molecules is a central technique in cell biology that continues to grow with the introduction of new fluorescent probes possessing unique properties and new methodologies for introducing them into cells [1]. In recent years, the unique spectral properties of semiconductor quantum dots (QDs) coupled with three significant advances are as follows: (1) new QD capping ligands for rendering the nanocrystals colloidally stable in biological media, (2) bioconjugation chemistries for the attachment of biomolecules to the QD surface, and (3) methods for facilitated QD delivery to cells [2–5], have driven their widespread use in biological applications in in vitro cellular labeling/imaging and sensing. In spite of their verified utility, concerns over the potential toxic effects of QD core materials on cellular

Xue-Long Sun (ed.), *Macro-Glycoligands: Methods and Protocols*, Methods in Molecular Biology, vol. 1367,
DOI 10.1007/978-1-4939-3130-9_9,

proliferation and homeostasis have persisted, leaving in question the suitability of QDs as alternatives for more traditional fluorescent materials (e.g., organic dyes, fluorescent proteins) for in vitro cellular applications [1]. One way of improving QD bioavailability has been incorporation of carbohydrates and biotin on the surface of the QDs, via a simple EDC coupling and RAFT polymerization method. This way of introducing carbohydrate and biotin onto QD surface is known to improve the solubility and biocompatibility and reduce toxicity of fluorescent nanocrystals. It is also able to maintain optical properties of the original QD nanocrystals [6].

The original QDs (yellow color) used in this study is a carboxy-functionalized in CdS-capped CdTe nanoparticle solution, with sodium as carboxy counter ion, and stabilized in aqueous solution by the carboxyl terminated polymer chains. The polymers improve their water solubility in basic conditions and provide the surface for further functionalization. In order to maximize the biological functions of the QDs, we have suggested two different strategies to prepare stable surface functionalized QDs with both biotin and carbohydrates moieties via 1-ethyl-3-(3-dimethylaminopropyl)carbodiimide (EDC) coupling [6]. Our motivation here is to incorporate carbohydrate and biotin moieties on the surface of QDs for improving biocompatibility.

The first approach is to modify the COOH groups with small biotin and carbohydrate molecules bearing amine groups. In this approach the biotin bearing terminal amino group was fabricated via a two-step synthesis to functionalize the carboxyl-capped QDs from D-biotin. The D-biotin was first activated by the NHS to form biotin-NHS and then reacted with ethylenediamine to obtain biotin-NH2 as shown in Fig. 1a. Similarly, the sugar bearing terminal amino group (sugar-NH_2) was synthesized by the reaction of D-(+)-gluconic acid δ-lactone with ethylenediamine hydrochloride as shown in Fig. 1b. The resultant biotin-NH_2 and carbohydrate bearing terminal amino group (sugar-NH_2) contains primary amino groups as our functionalization targeted moiety to react with the activated carboxylic groups on the QDs [7, 8].

In the second approach, the QD surface was modified with biotinylated glycopolymers bearing biotin and carbohydrate as pendent moieties. The biotin monomer, BAEMA was prepared by the reaction of biotin-NHS with AEMA. Subsequently, amine-based monomer, 2-aminoethyl methacrylamide hydrochloride (AEMA), carbohydrate-containing monomer, 2-gluconamidoethyl methacrylamide (GAEMA) and biotin containing monomer, BAEMA, were copolymerized statistically via RAFT method (Fig. 2). The obtained biotinylated glycopolymer [P(GAEMA-st-AEMA-*st*-BAEMA)] possesses end amino groups for realizing the possibility to link the polymer chain onto the surface of original QDs. Both synthetic routes are considered in details below. Consequently, original QDs surface

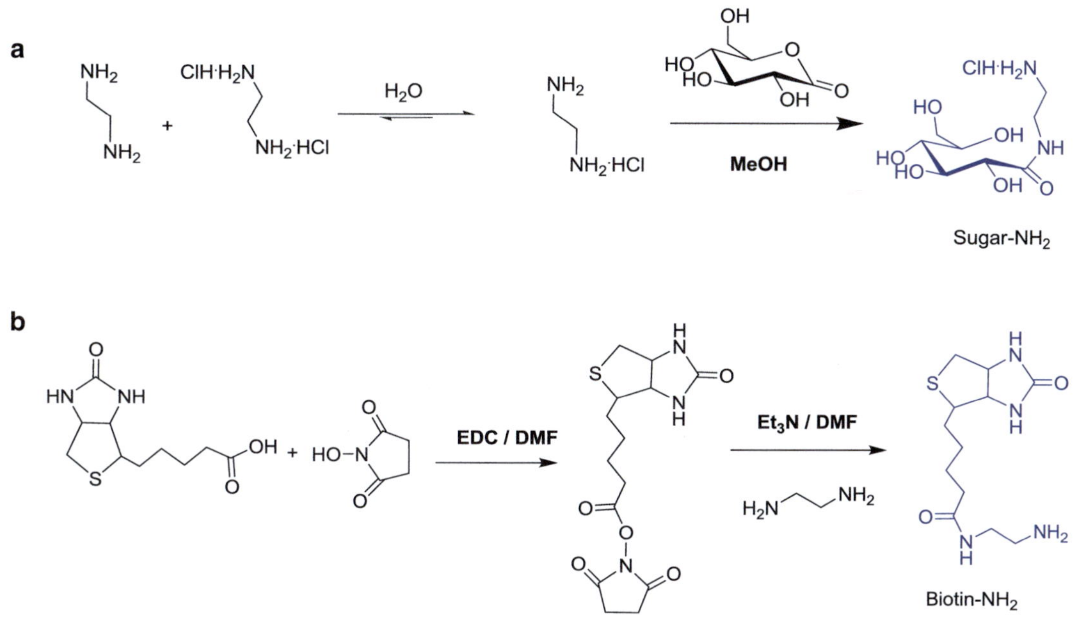

Fig. 1 Preparation of biotin-NH_2 and sugar-NH_2 (*see* **Note 1**)

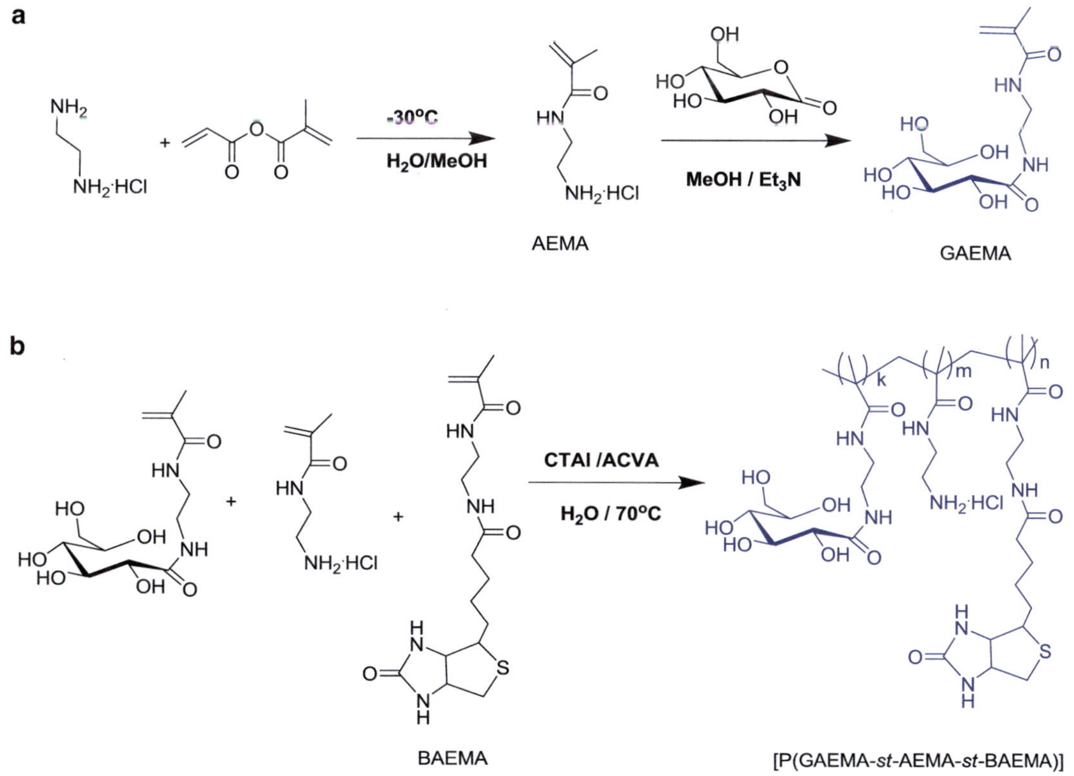

Fig. 2 Preparation of biotinylated glycopolymer. (**a**) An approach suggests synthesis of alkyl amine pendent as the primary amine fragment followed by incorporation of glyco monomer motif [7, 8] (*see* **Notes 2–4**). (**b**) Synthetic approach leading to glycopolymer with several biotins and primary amine motifs [6]

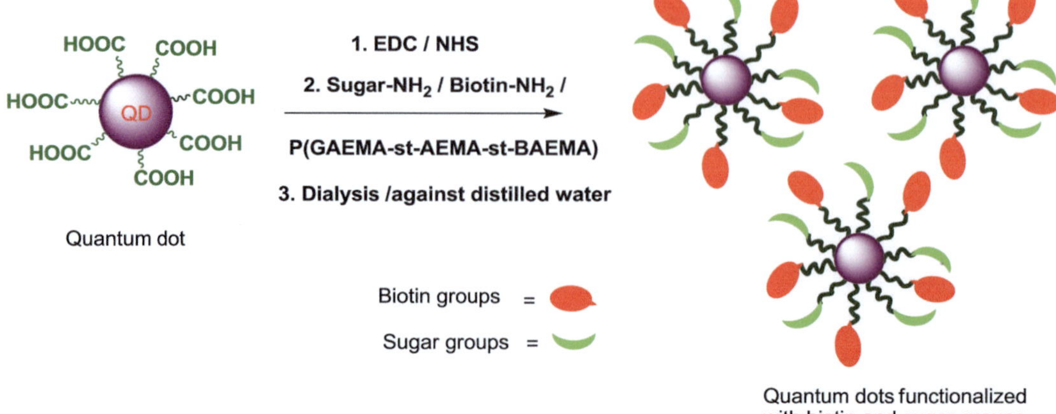

Fig. 3 Schematic illustration of the functionalized QDs fabrication with biotin and sugar or glycopolymer

were functionalized by dispersing QDs nanoparticles in basic aqueous solution and reacted with the biotinylated glycopolymer or the mixture of biotin-NH_2 and sugar-NH_2 as shown in Fig. 3. Also we describe here the bioavailability of the biotin ligands on the surface of functionalized QDs is quantified.The detailed synthetic routes are considered below.

2 Materials

2.1 Chemicals

1. 4′-Azobis(4-cyanovaleric acid) (ACVA, 97 %) (Acros Organics).
2. D-Biotin.
3. 1, 3-Dicyclohexylcarbodiimide (DCC).
4. *N*, *N*′-Dimethylformamide (DMF).
5. 1, 4-Dioxane.
   1-Ethyl-3-[3-dimethylaminopropyl] carbodiimide hydrochloride (EDC).
6. Ethylenediamine.
7. Ethylenediamine dihydrochloride.
8. D-(+)-Gluconic acid δ-lactone.
9. Isopropanol.
10. Hydroquinone.
11. *N*-Hydroxysuccinimide (NHS).
12. *S*, *S*′-bis(R, R′-dimethyl-R″-acetic acid) trithiocarbonate (CTA1) was synthesized as described in ref. [9].
13. Methanol.

14. Quantum dots (yellow color, carboxy-functionalized CdS capped CdTe nanoparticle solution, sodium as carboxyl counterion; 20 mg/mL in water, and emission maxima wavelength and full-width at half-maximum (FWHM) are 563 and 90 nm at 350 nm of excitation, respectively) with surface carboxylic groups (Northern Nanotechnologies Inc.).

2.2 Buffer and Other Solutions

1. Doubly distilled deionized water was used in all experiments.
2. PBS buffer solution (pH 7.4) 1.0 M.
3. Buffer solution for GCP: sodium acetate 0.5 M/acetic acid.
4. Six near-monodisperse PEO standards (Mp) 1010–101,200 g/mol).

2.3 NMR Spectroscopy

^{1}H NMR spectra of the monomers and polymers are recorded on a Varian 200 MHz instrument.

2.4 Chromatography

2.4.1 Preparative GPC

1. Aqueous gel permeation chromatography analysis is performed on a Viscotek Instrument using acetate/acetic acid buffer as eluent, two Waters WAT011545 columns, and a flow rate of 1.0 mL/min.

2.4.2 Analytical GPC

1. Six near-monodisperse PEO standards (Mp) 1010–101,200 g mol^{-1}) are used for calibration.
2. The products are analyzed by an Agilent HPLC 1100 interfaced with an Electro-Spray Ionization Agilent Mass Spectrometer Model 6120 with a Chemstation data system LCMSD B.03.01.

2.5 Dynamic Light Scattering

1. Dynamic light scattering (DLS) is performed at room temperature using a Viscotek DLS instrument with an He–Ne laser at a wavelength of 632 nm and a Pelletier temperature controller.
2. The surface functionalized quantum dot aqueous solutions (20 nM) are filtered through Millipore membranes (0.45 μm pore size).
3. The data is recorded with OmniSize Software.

2.6 Fluorescence Measurements

1. Fluorescence measurements are performed on an OLISRSM-1000 (Desa rapid-scanning monochromator spectrophotometry system).
2. The spectra are recorded in the wavelength range of 500–650 nm upon excitation at 490 nm; a 1.00 cm path length rectangular quartz cell was used for this study.
3. Very dilute solutions of surface functionalized quantum dots (20 nM) are used in the experiment.

3 Methods

The EDC coupling approach and two different strategies were used to prepare stable surface functionalized QDs with both biotin and carbohydrates moieties. The first strategy is that QD surface was functionalized with two monomers: biotin bearing terminal amino group (biotine-NH_2) and sugar bearing terminal amino group (sugar-NH_2). The biotin bearing terminal amino group was fabricated via a two-step synthesis to functionalize the carboxyl capped QDs from D-biotin. The D-biotin was first activated by the NHS to form biotin-NHS and then reacted with ethylenediamine to obtain biotin-NH_2 as shown in Figure 1a (*see* **Note 1**). The product was characterized by ^1H NMR in *d*6-DMSO solvent and mass spectrometry to confirm the formation of the desired product. Similarly, the sugar bearing terminal amino group (sugar-NH2) was also synthesized by the reaction of D-(+)-gluconic acid-δ-lactone with ethylenediamine as shown in Figure 1b. In this case dihydrochloride salt of aliphatic primary amine and aliphatic primary amine were mixed together to give partially deprotonated hydrochloride salt of aliphatic primary amine in situ, based on the equilibrium between aliphatic primary amine and its dihydrochloride salt. A sticky yellow product was obtained in high yield (~90 %), and the chemical structure was confirmed by mass spectrometry.

The second strategy is the formulation of a biotinylated glycopolymer prepared via RAFT method for QDs surface functionalization. This was achieved by RAFT polymerization of biotin monomer, primary aminoalkyl methacrylamide monomer (AEMA) and sugar monomer (GAEMA). The biotin containing monomer, BAEMA, was prepared by the reaction of biotin-NHS with AEMA. Prior to BAEMA synthesis, primary aminoalkyl methacrylamide monomer (AEMA) was synthesized by reacting dihydrochloride salt of aliphatic primary amine with its aliphatic primary amine at low temperature (−30 °C) followed by addition of methacrylic anhydride. Based on the equilibrium between aliphatic primary amine and its dihydrochloride salt, the formation of dimethacrylamide, one of the main side reactions, was minimized. The sugar monomer, GAEMA, was also readily synthesized in methanol at room temperature by the reaction of AEMA with D-gluconolactone. Ring-opening of the D-gluconolactone only occurs in the presence of the triethylamine, which reacts with the HCl to generate the free primary amine in situ [10]. This synthetic approach leads to monofunctional GAEMA, with essentially no side reactions. The monomers BEAMA, AEMA, and GAEMA were polymerized statistically via RAFT in water at 70 °C. Polymerization was carried out using CTA1 as the chain transfer agent and ACVA as the initiator. The ratio of CTA to initiator was set to 5. The polymer obtained by this method has narrower molecular mass distribution with a very low polydispersity index (1.19), which has a M_n ~12,900 g/mol. The final statistical glycopolymer obtained was P(GAEMA$_{36}$-*st*-AEMA$_{24}$-*st*-BAEMA$_{4.4}$) from the ^{1}H NMR and HABA/avidin binding assay results.

Finally, we describe a method leading to a biologically active biotinylated glyco-functionalized QD. This was synthesized by dispersing original QDs in basic aqueous solution, but they rapidly lost fluorescence below pH 6–6.5. The carboxylic groups on the surface of the QDs are activated using the NHS/EDC coupling method when the pH of the QDs solution was fixed at 7.4 using PBS. Subsequently, the QD is reacted with the biotinylated glycopolymer or the mixture of biotin-NH_2 and sugar-NH_2 (Fig. 3). The primary amino group on the statistical copolymer or biotin and sugar reagents was targeted to react with the activated carboxylic groups on the QDs.

Fluorescence spectra were recorded on the original QDs, biotin- and sugar-functionalized QDs, and biotinylated glycopolymer-functionalized QDs. Both surface-functionalized QDs still show strong fluorescence indicating that the surface modification did not have a significant effect in the optical properties of the original QDs (Fig. 4d). The size distribution of original QDs, biotin and sugar functionalized QDs, and biotinylated glycopolymer-functionalized QDs were studied by dynamic light scattering. The size distributions of both surface functionalized QDs are relatively narrow (less than 0.15). The size of biotin and sugar functionalized QDs are similar to that of original QDs (~6.0 nm), which further confirms that surface functionalization has no influence on the physical states of original QDs (Fig. 4a, b). The glycopolymer-functionalized QDs revealed an increase in size (10–11 nm), which confirms the successful functionalization of the glycopolymer to the QDs (Fig. 4c).

We also describe a method of calculating the availability of biotin on the surface of the QD quantitatively via HABA/avidin binding assay. This was achieved by quantitatively monitoring the decrease in UV absorbance of HABA/avidin solution at a wavelength of 500 nm after the displacement of HABA by available biotin. HABA complexes with avidin in water and the aqueous solution have a maximum absorbance at 500 nm. Upon addition of biotin or biotinylated reagents to the HABA/avidin solution, HABA is displaced quantitatively by available biotin, as avidin's affinity for biotin ($Kd = 10^{-15}$ M) is much higher than that of HABA ($Kd = 10^{-6}$ M). However, upon addition of the functionalized QDs, the absorbance value at 500 nm of (Fig. 5) HABA/avidin clearly decreases, suggesting that HABA is displaced from the HABA/avidin complex by available biotin presented on the QD surface.

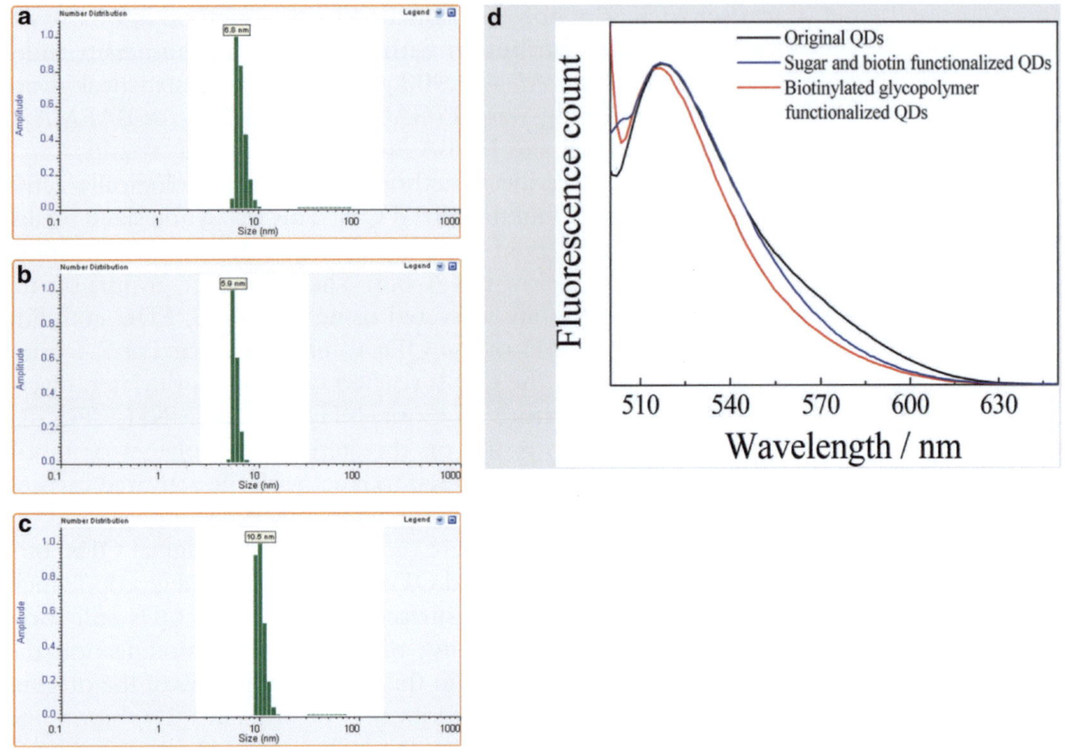

Fig. 4 (**a**) DLS number average distribution of original QDs, (**b**) biotin- and sugar-functionalized QDs, (**c**) biotinylated glycopolymer-functionalized QDs, and (**d**) emission spectra of original QDs, biotin- and sugar-functionalized QDs, and biotinylated glycopolymer-functionalized QDs, where the excitation wavelength is at 490 nm [6]

The amount of biotin in each biotinylated QD solution could be calculated quantitatively by the following formula: $\Delta A500 = 0.9\ A_{HABA/avidin} + A_{QDs} - A_{HABA/avidin+QDs}$

μmol biotin/mL = $10(\Delta A_{500}/34)$

Where the dilution factor of HABA/avidin is 0.9, the mM extinction coefficient of biotin at 500 nm is 34, and the dilution factor of QDs is 10.

The absorbance of aqueous solution of QDs before or after surface functionalization is about 0.238 at 500 nm after surface functionalization of QDs using polymer chains. So the total decreased absorbance of the HABA/avidin complex after the addition of surface-functionalized QDs should be considered the initial absorbance of QDs. However, the total absorbance after mixing calculated by the above formula is 0.11. The amounts of available biotin per mL of QDs solution were thus calculated to be 1.29 μmol and 0.93 μmol for the glycopolymer-functionalized QDs and biotin- and sugar-functionalized QDs, respectively. But according to the concentration of original QDs, which is about 0.8 μM, the quantity of available biotin per QD could be calculated to be

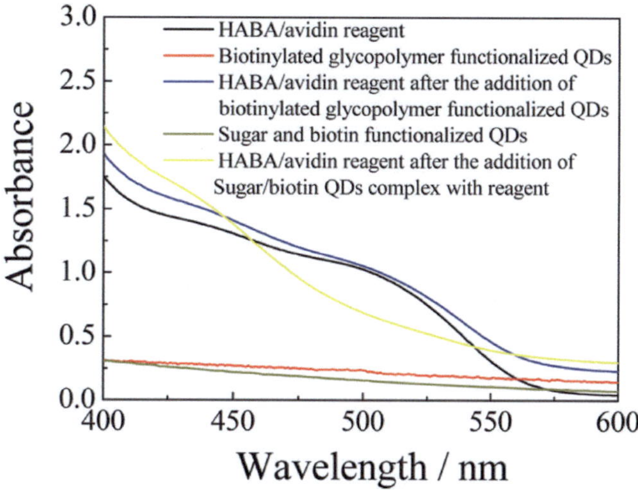

Fig. 5 UV–Vis spectra of the HABA/avidin reagent before and after the addition of surface-functionalized QDs, biotinylated glycopolymer QDs and sugar- and biotin-functionalized QDs [6] where the QDs concentrations were 0.02 μM and 0.1 μM for biotinylated glycopolymer QDs and sugar- and biotin-functionalized QDs, respectively

~1600 and ~1100 for biotinylated glycopolymer-functionalized QDs and biotin- and sugar-functionalized QDs, respectively.

4 Preparation of Biotin and Sugar Monomers

4.1 Synthesis of Biotinyl-N-Hydroxysuccinimide Ester (Biotin-NHS)

1. Dissolve D-biotin (2.00 g, 8.19 mmol) and *N*-hydroxysuccinimide (0.94 g, 8.19 mmol) into hot anhydrous DMF (60 mL, 70 °C) in a 100 mL round-bottom flask with stirring.
2. Add 1,3-dicyclohexylcarbodiimide (DCC, 2.19 g, 10.65 mmol), and stir the solution overnight at room temperature.
3. Filter off the formed dicyclohexylurea; evaporate most of the solvents, and precipitate the residue into excess ether.
4. Collect the white precipitate by filtration and wash with isopropanol three times to give a white powder, and recrystallize in isopropanol.
5. The final yield of the product is 85 % and its structure is confirmed via ^{1}H NMR spectroscopy.

4.1.1 Synthesis of N-(2-Aminoethyl) Biotinamide (Biotin-NH_2)

1. Add ethylenediamine (150 μL, 2.2 mmol), triethylamine (1.5 mL), and DMF (10 mL) into a round-bottom 50 mL flask fitted with a stirrer.

2. Add a DMF solution (5 mL) of biotin-NHS (0.75 g, 2.2 mmol) drop wise at room temperature under nitrogen atmosphere and stir overnight.
3. Filter off the formed precipitate, evaporate most of the solvents, and precipitate the residue into excess hexane.
4. Collect the white precipitate by filtration and wash with isopropanol three times to give a white powder. The final yield of the product is 70 %.

4.1.2 Synthesis of 2-Aminoethyl Methacrylamide Hydrochloride (AEMA)

1. Add an ethylenediamine dihydrochloride solution (10 g, 75 mmol, in 200 mL distilled water) to a solution of ethylenediamine (11 mL, 0.16 mol) in a three-necked round-bottom flask fitted with a head stirrer and a thermometer.
2. After 1 h of stirring, ad MeOH (230 mL) and cool the mixture down to −30 °C.
3. Mix methacrylic anhydride (22 mL, 0.15 mol) and hydroquinone first with methanol (30 mL) and subsequently add to the mixture above.
4. After the complete addition of the methacrylic anhydride, maintain the solution at −30 °C for 75 min and then add hydrochloric acid (24 mL) and the pH of the solution is recorded and the solution is maintained at a pH of ~1 overnight (*see* **Notes 2** and **3**) .
5. Finally, remove the solvent under vacuum to afford the crude creamy product and wash with acetone.
6. Extract the product by hot 2-propanol and followed by crystallization in cold 2-propanol. Yield 60 %.

4.1.3 Synthesis of N-(2-Aminoethyl) Gluconamide Hydrochloride (Sugar-NH_2)

1. Add ethylenediamine (5.5 mL, 0.08 mol) to an ethylenediamine dihydrochloride solution (12.76 g, 96 mmol, in 100 mL distilled water) in a round-bottom flask fitted with a stirrer.
2. After 1 h of stirring, add aqueous solution (70 mL) of D-(+)-gluconic acid δ-lactone (14.25 g, 0.08 mol) dropwise into the above solution cooled in an ice bath condition.
3. Stir the mixture at room temperature overnight
4. Remove the solvent under vacuum to afford the crude product and wash with isopropanol twice.
5. Dissolve the product into methanol and precipitate into isopropanol twice to afford the viscous product and dry under vacuum. Yield 90 %.

4.1.4 Synthesis of Biotinyl-2-Aminoethyl Methacrylamide Hydrochloride (BAEMA)

1. Dissolve biotin-NHS (0.85 g, 2.5 mmol), hydroquinone (0.05 g), triethylamine (0.7 mL), and 2-aminoethyl methacrylamide hydrochloride (AEMA, 0.41 g, 2.5 mmol) in 15 mL anhydrous DMF (*see* **Note 6**).
2. Stir the solution overnight at room temperature,

3. Filter off the solid formed off and remove DMF with evaporator under reduced pressure.
4. Precipitate the residue into ether and washed with isopropanol three times to afford the powder and dry overnight under vacuum. Yield 70 %.

4.1.5 Synthesis of Gluconamidoethyl Methacrylamide Hydrochloride (GAEMA)

1. Dissolve D-(+)-gluconic acid δ-lactone (1.77 g, 0.08 mol) and hydroquinone (0.05 g) in methanol (50.0 mL) at 50 °C and then cool it to room temperature (*see* **Note 6**).
2. Add 2-aminoethyl methacrylamide hydrochloride (AEMA) (2.0 g, 0.012 mmol) and triethylamine (10 mL).
3. Stir the mixture overnight at room temperature.
4. Concentrate the reaction solution by rotary evaporation to afford a residue.
5. Precipitate the residue into isopropanol.
6. Collect the white solid formed by filtration, wash it with isopropanol three times, and dry under vacuum. Yield 95 %.

4.2 Preparation of P(GAEMA-Stat-AEMA-Stat-BAEMA) Copolymer via RAFT Polymerization

1. In a 10 mL flask, dissolve GAEMA (0.5 g, 1.46 mmol) and BAEMA (0.21 g, 0.54 mmol) were dissolved in distilled water (3 mL) at 40 °C and then cool to room temperature.
2. Add AEMA (0.18 g, 1.09 mmol, dissolved in 1.0 mL water), the mixture of ACVA (2.0 mg, 0.007 mmol) and CTA1 (10 mg, 0.035 mmol) dissolved into 2 mL of dioxane into the above solution (*see* **Notes 4** and **5**).
3. Degas the solution via three freeze-pump-thaw cycles.
4. Place the flask in an oil bath preheated at 70 °C to start the polymerization for 18 h.
5. After 18 h, place the flask into liquid nitrogen to quench the polymerization.
6. Dilute the mixture with water (5.0 mL) and then precipitate into an excess of isopropanol. Repeated this purification cycle twice.
7. Collect the obtained slightly yellow powder and dry in a vacuum oven overnight at room temperature. Yield 67 %.

4.3 Synthesis of Biotinylated Glyco-Functionalized Quantum Dots via EDC Coupling

1. Add Quantum dots solution (1 mL) to 1.0 M PBS buffer solution (pH 7.4, 4 mL) of EDC (2.6 mg) and NHS (1.5 mg) (*see* **Note 7**).
2. Stir the reaction mixture for 30 min at room temperature and then add P(GAEMA-*st*-AEMA-*st*-BAEMA) copolymer (10.0 mg in 35 mL of PBS buffer solution) or the mixture of sugar-NH_2 (2.6 mg, dissolved in 34 mL of PBS buffer solution) and biotin-NH_2 (1.1 mg in 1 mL of DMF) to the above PBS buffer solution.

3. After stirring the reaction mixture overnight at room temperature, dialyze the mixture against distilled water for 2 days using the dialysis membrane molecular weight cutoff (MWCO) of 12,000–14,000 and freeze-dry the desired QDs under vacuum for 2 days.

4.4 Quantification of Amount of Biotin on the QDs Surface Using HABA/Avidin Binding Assay

1. Before the samples testing, filter the solutions through Millipore membranes (0.45 μm pore size) and then concentrate or dilute to the appropriate concentration.
2. In a 1.0 mL cuvette, pipet 900 μL HABA/avidin reagent (0.08 μmole/mL biotin).
3. Read A_{500} and add 100 μL sample, mix by inversion, and then read A_{500}.
4. Dilute 100 μL sample with 900 μL water or diluent as a blank [11, 12].
5. Record the change of absorbance at 500 nm of HABA/avidin and after HABA/avidin adding surface functionalized quantum dots (should be at 0.1–0.4).

5 Notes

1. In order to minimize side reactions in the fabrication process of biotin-NH_2, the DMF solution of BNHS is added drop wise into the flask containing ethylenediamine to avoid any possible side reactions.
2. In the preparation of AEMA, the reaction can also be carried out at room temperature instead at −30 °C but the yield of the final product obtained is slightly lower, that is, yield ~50%.
3. Thus the AEMA reaction is conducted at low temperature, −30 °C, mainly to reduce the reaction rate, which also helps to minimize side reactions, for instance, the formation of dimethacrylamide, and offers an opportunity to terminate the reaction in a proper time.
4. The AEMA monomer is used in the protonated form during polymerization to prevent any side reactions of the primary amino group to the chain transfer groups.
5. Polymerization is carried out using CTA1 as the chain transfer agent and ACVA as the initiator and the ratio of CTA to initiator was set to 5.
6. Small amount of hydroquinone is always added in the preparation of BAEMA and GAEMA to avoid self polymerization of monomers.
7. To prevent possible flocculation of the QDs, the coupling reaction is carried out under high dilution to decrease the possibility of side reactions.

References

1. Bradburne CE, Delehanty JB, Gemmill KB, Mei BC, Mattoussi H, Susumu K et al (2013) Cytotoxicity of quantum dots used for in vitro cellular labeling: role of QD surface ligand, delivery modality, cell type, and direct comparison to organic fluorophores. Bioconjugate Chem 24:1570–1583
2. Algar WR, Susumu K, Delehanty JB, Medintz IL (2011) Semiconductor quantum dots in bioanalysis: crossing the valley of death. Anal Chem 83:8826–8837
3. Delehanty JB, Mattoussi H, Medintz IL (2009) Delivering quantum dots into cells: strategies, progress and remaining issues. Anal Bioanal Chem 393:1091–1105
4. Delehanty JB, Susumu K, Manthe RL, Algar WR, Medintz IL (2012) Active cellular sensing with quantum dots: transitioning from research tool to reality; a review. Anal Chim Acta 750:63–81
5. Algar WR, Prasuhn DE, Stewart MH, Jennings TL, Blanco-Canosa JB, Dawson PE, Medintz IL (2011) The controlled display of biomolecules on nanoparticles: a challenge suited to bioorthogonal chemistry. Bioconjugate Chem 22:825–858
6. Jiang X, Ahmed M, Deng Z, Narain R (2009) Biotinylated glyco-functionalized quantum dots: synthesis, characterization, and cytotoxicity studies. Bioconjugate Chem 20:994–1001
7. Deng Z, Ahmed M, Narain R (2009) Novel well-defined glycopolymers synthesized via the reversible addition fragmentation chain transfer process in aqueous media. J Pol Sci A Pol Chem 47:614–627
8. Deng Z, BouchéKif H, Babooram K, Housni A, Choytun N, Narain R (2008) Facile Synthesis of controlled-structure primary amine-based methacrylamide polymers via the reversible addition-fragmentation chain transfer process. J Pol Sci A Pol Chem 46:4984–4996
9. Lai JT, Filla D, Shea R (2002) Functional polymers from novel carboxyl-terminated trithiocarbonates as highly efficient RAFT agents. Macromolecules 35:6754–6756
10. Narain R, Armes SP (2002) Synthesis of low polydispersity, controlled-structure sugar methacrylate polymers under mild conditions without protecting group chemistry. Chem Commun 23:2776–2777
11. Green NM (1965) A spectrophotometric assay for avidin and biotin based on binding of dyes by avidin. Biochem J 94:23c–24c
12. Green NM (1970) Spectrophotometric determination of avidin and biotin. Methods Enzymol 18:418–424

Chapter 10

Surface Modification of Polydivinylbenzene Microspheres with a Fluorinated Glycopolymer Using Thiol-Halogen Click Chemistry

Wentao Song and Anthony M. Granville

Abstract

Distillation-precipitation polymerization of divinylbenzene was applied to obtain uniform-sized polymeric microspheres. The microspheres were then modified with polypentafluorostyrene chains utilizing surface-initiated atom transfer radical polymerization techniques. The hydrophobic fluoropolymer-coated microsphere was then converted to a hydrophilic biopolymer by performing thiol-halogen click chemistry between polypentafluorostyrene and 1-thio-β-D-glucose sodium salt. The semi-fluorinated glycopolymer showed good binding ability with Concanavalin A as determined by confocal microscopy and turbidity experiments.

Key words Polymeric microspheres, Surface modification, Glycopolymers, Atom transfer radical polymerization, Protein binding

1 Introduction

In the field of controlled drug delivery systems and biomedical assay devices, increased attention has been paid to synthetic polymers substituted with pendant carbohydrate groups as biological recognition units [1–5]. Thiol-halogen nucleophilic substitution "click" chemistry allows for the formation of these glycopolymers from fluorinated aromatic polymers using mild and efficient reaction conditions [6]. In 2001, Sharpless et al. [7] introduced the new concept of "click" chemistry, which represents a series of highly selective, simple orthogonal reactions that do not yield side products and that give heteroatom-linked molecular systems with high efficiency under a variety of mild conditions.

Thiols, which are prototypical soft nucleophiles, can participate in these types of substitution reactions with reactive substrates bearing readily displaceable leaving groups, such as halogens. The ease with which these reactions occur resulted in thiol–halogen

Xue-Long Sun (ed.), *Macro-Glycoligands: Methods and Protocols*, Methods in Molecular Biology, vol. 1367,
DOI 10.1007/978-1-4939-3130-9_10, © Springer Science+Business Media New York 2016

nucleophilic substitution reactions being touted as "click" reactions [1, 8, 9]. The halogenated salts produced can be readily removed as precipitants in a facile purification procedure. The work of Rosen et al. [8] demonstrated the thiol functionality of 2-mercaptoethanol added exclusively to halogen end-functionalized polymers and other halogenated species due to its high relative nucleophilicity compared to alcohols. This result provides a powerful conduit for selective end-functionalization of polymer chain ends and attests to the orthogonality of the thiol–halogen "click" reaction; that is, the halogen is displaced by thiols even in the presence of a large excess concentration of alcohol groups.

Of particular interest is the thiol-halogen "click" reaction between thiols and the labile fluorines in the para position of pentafluorophenyl groups. Recently Samaroo et al. [9] showed that porphyrinoid macrocyclics with four pentafluorophenyl groups undergo rapid and nearly quantitative reactions with a cadre of aromatic and aliphatic thiols exhibiting high yields utilizing various mild bases. They demonstrated the potential of using such "click" reactions to effectively modify biologically important cores to enhance solubility, intermolecular binding, aggregation and transport in biological media. By a very similar process, Becer et al. [1] prepared glycopolymers via nucleophilic substitution of the p-fluorine group of a pentafluorostyrene homopolymer and a styrene–pentafluorostyrene block copolymer with a glucosylthiol. ^{19}F NMR kinetics indicated that under mild conditions with a trialkyl amine base, the one product substitution reactions were essentially quantitative on the span of a little over one hour. It was speculated that the metal free thiol–fluorine "click" displacement reaction could be utilized to produce functionalized nanoparticles for coating material components for body implants [1]. This would be of particular interest, as fluorinated compounds are known for their antifouling properties due to their low surface and adhesive energies [10, 11].

By coupling fluorinated glycopolymers with polymeric microspheres, it is possible that these materials could be useful for lectin affinity chromatography to separate proteins and drug delivery systems [12–14]. Employing distillation-precipitation to synthesize polymeric microspheres is a simple and elegant strategy to obtain the carrier particles with a functional surface [15]. Controlled radical polymerization, such as atom transfer radical polymerization (ATRP) and reversible addition-fragmentation chain transfer (RAFT) polymerization, allow for facile surface modification and eventual surface polymerization of monomers exhibiting biological mimicking capabilities. In our research, ATRP was carried out in a "grafting from" approach to obtain well-defined polypentafluorostyrene (PPFS) modified polymeric microspheres (Fig. 1). These fluorinated materials were then converted to fluorinated glycopolymers using a thioglucose salt and thiol-halogen "click" chemistry without the need for any further deprotection chemistry.

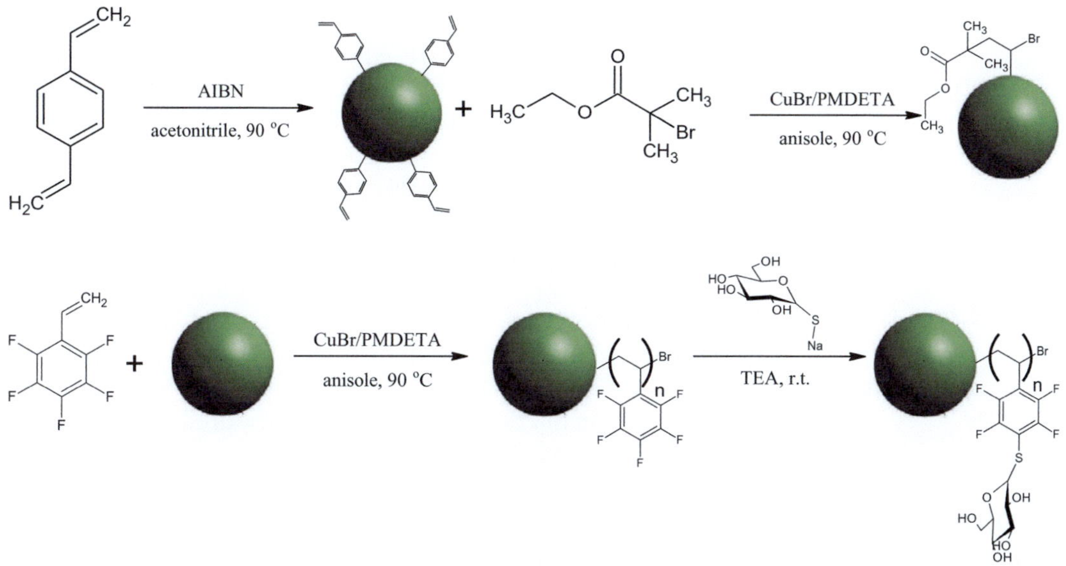

Fig. 1 Size-exclusion chromatography (SEC) for PPFS and Gluc-PPFS samples generated from solution ATRP

2 Materials

2.1 Chemicals

1. 2,2′-Azobis(2-methylpropionitrile) (AIBN, 98 %) is recrystallized twice from methanol prior to use.
2. Concanavalin A (Con A).
3. Concanavalin A, FITC conjugate (FITC-Con A)
4. Copper(I) bromide (97 %) is purified to remove oxides (*see* **Note 1**).
5. Divinylbenzene (DVB80, 80 % isomers) (Aldrich) is passed through a column of aluminium oxide prior to use (*see* **Note 2**).
6. Ethyl 2-bromoisobutyrate (E2BriB, 98 %).
7. 10 mM (pH 7.5) 4-(2-Hydroxyethyl)-1-piperazineethanesulfonic acid (HEPES, Fluka) buffer solution.
8. *N,N,N′,N′,N″*-Pentamethyldiethylenetriamine (PMDETA, 97 %).
9. Pentafluorostyrene (PFS, 99 %) (Oakwood Products) is passed through a column of aluminium oxide prior to use (*see* **Note 2**).
10. 1-Thio-β-D-glucose sodium salt.

2.2 ^1H and ^{19}F NMR

NMR spectra were obtained on a Bruker DPX 300 MHz spectrometer in either deuterated chloroform (CDCl3) or deuterated dimethyl sulfoxide (d-DMSO) solvents.

2.3 Size-Exclusion Chromatography

Size-exclusion chromatography (SEC) analysis of the subsequent glycopolymers were performed in N,N-dimethylacetamide (DMAc) (0.03 % w/v LiBr, 0.05 % BHT) at 50 °C (1 mL/min flow rate) using a Shimadzu modular system comprising a DGU-12A solvent degasser, SIL-10 AD auto-injector, an LC-10AT pump, a CTO-10A column oven, and an RID-10A refractive index detector. The system was equipped with a 5.0 μm bead-size guard column (50 × 7.8 mm) followed by four 300 × 7.8 mm linear Phenomenex columns (105, 104, 103, and 500 Å). The calibration curve was generated with narrow polydispersity polystyrene standards ranging from 500 to 106 g/mol.

2.4 ATR-FTIR

ATR-FTIR absorption spectra were recorded on a Bruker VECTOR-22 FTIR spectrometer.

2.5 SEM

Scanning electron microscopy (SEM) was measured using a Hitachi S4500, with an accelerating voltage of 5 kV.

2.6 Dynamic Light Scattering

Particle size distribution and zeta potential were measured by a Malvern Zetasizer Nano ZS equipped with a 4 mW He-Ne laser ($\lambda = 632.8$ nm). The instrument was calibrated with titanium oxide standard (RI = 2.40, absorption = 0.01). At least three measurements were performed on each sample, and each measurement consisted of 12–14 scans. The sample was dispersed at low concentration in deionized water (RI = 1.33, absorption = 0.01) by sonication prior to analysis.

2.7 Confocal Microscope

Confocal microscopy experiments were performed on a Zeiss Spinning disk microscope Observer X.1 with objective: 63× water. The spinning disk unit is a Yokogawa CSU-X1. The fluorescein was imaged using a BP 525/50 filter. NA1.3 Argon ion laser, 488 nm line, EMCCD camera: Photometrics QuantEM 512SC.

3 Methods

A method for generating semifluorinated glycopolymer modified polymer particles for protein binding is described. The technique involves the distillation-precipitation polymerization to generate uniform microspheres, the controlled polymerization of PFS, as well as the thiol click modification of the PPFS to a glycopolymer brush and binding studies. The ATRP kinetics of PFS in anisole using a Cu(I)Br and PMDETA system was found to give linear polymer chain growth with respect to conversion with a low dispersity. PPFS chains ($M_n = 17{,}000$ g/mol and $Đ = 1.04$) were obtained after 16 h polymerization time, which was used to generate the polymer brushes on pDVB microspheres (Fig. 1). The high-molecular-weight shoulder observed for the PPFS sample is indicative of termination due to coupling as a result of the high monomer conversion and is not observed when the conversion is kept below 90 %.

Thiols are well known as soft nucleophiles in comparison to primary amines or alcohols, hence displaying higher reactivity in nucleophilic substitution reactions. The reaction occurs with quantitative yields under ambient conditions without any need for a metal catalyst. Therefore, the synthesized polypentafluorostyrene homopolymer (1 equiv. with respect to the pentafluorostyrene units) and 1-thio-β-D-glucose sodium salt (1.2 equiv.) were dissolved in DMF, in the presence of triethylamine (3 equiv.) as a base catalyst, and allowed to react at room temperature for 6 h. Afterwards, the solution was precipitated into cold methanol to result in a white precipitate with an isolated yield of 98 %. This yield is determined using ^{19}F NMR, which typically shows a shift in the peaks as well as a loss of the *para*-fluorine of PPFS after glucosylation.

Concanavalin A (Con A) has been shown to exhibit a binding affinity for glucose moieties as well as being an activator for cellular signalling events, such as cell adhesion, proliferation, and survival [12]. Figure 2 shows the results for the turbidimeteric assay with Con A on the glucosylated fluoropolymer (Gluc-PPFS) in the HEPES buffer solution. As can be seen, the combined solution shows a quick initial decrease in transmission at 420 nm in the first 5 min and reaching a plateau after 20 min, which indicates a strong binding affinity to the fluorinated glycopolymer. This activity is

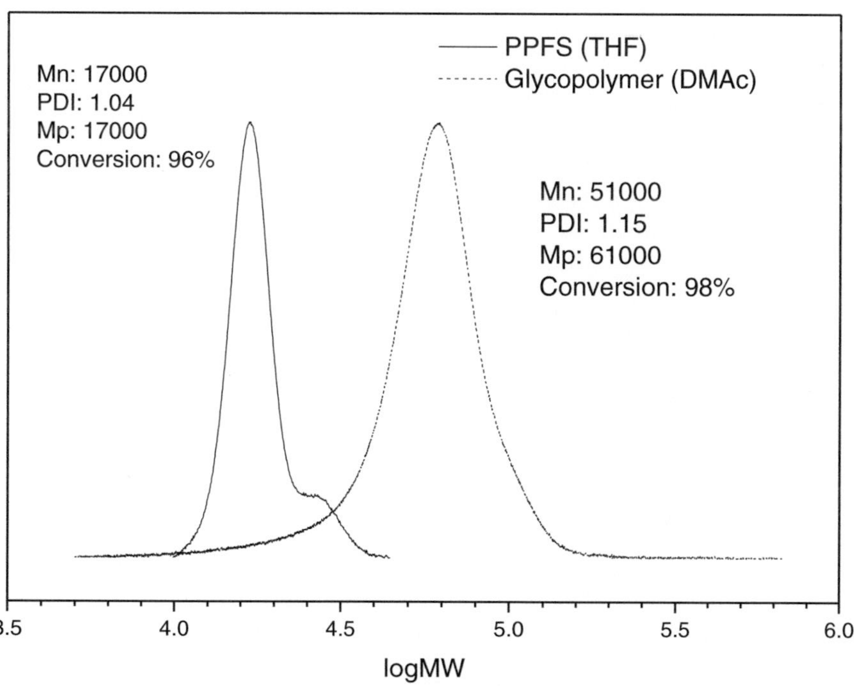

Fig. 2 Turbidity experiments showing the decreased transmittance, at 420 nm, with respect to time of the solution due to the binding and aggregation/precipitation of the polymer with the lectin

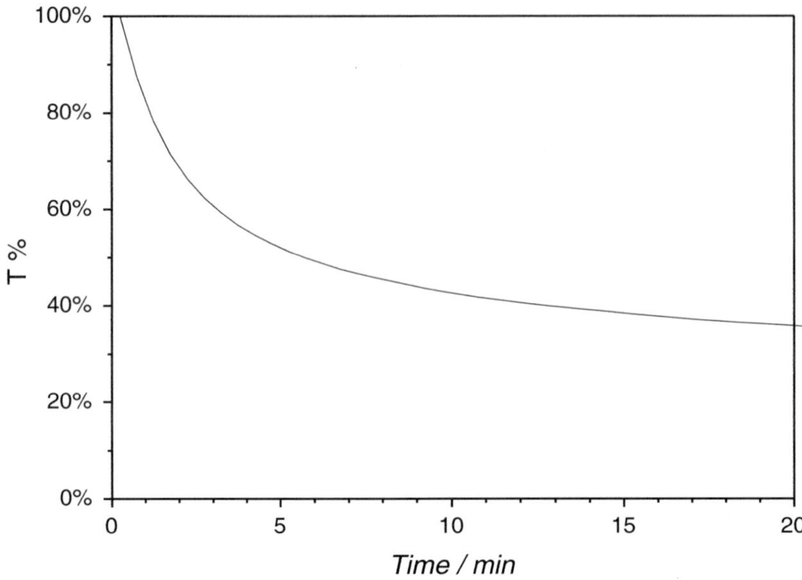

Fig. 3 Synthesis scheme for generating pDVB microspheres, modifying the surface with an ATRP initiator, generating fluorinated polymer brushes from the surface (PPFS-pDVB), and converting them to glycopolymer surfaces (Gluc-PPFS-pDVB)

similar to those observed for non-fluorinated polymer systems and typical for polymers which exhibit multiple binding sites to proteins. This same system was then extended to polymerize microsphere surfaces.

The whole strategy for the synthesis and modification of polydivinylbenzene microspheres is depicted in Fig. 3. The particle size and size distribution were determined by dynamic light scattering (DLS) and scanning electron microscopy (SEM, Fig. 4) and shown to be 1.5 μm diameter in size. These particles were pre-treated with an ATRP initiator so as to generate a surface initiated-ATRP particle surface. pDVB microspheres, copper bromide, PMDETA, and ethyl α-bromoisobutyrate were reacted in anisole at 90 °C for 8 h to ensure the initiator had completely reacted with vinyl groups on the surface. The polymeric particles were dried in a vacuum oven over night after filtering and washing the microspheres. The surface-modified pDVB microspheres were then used for polymerizing pentafluorostyrene via ATRP under similar conditions to the PPFS homopolymerization reactions (90 °C for 16 h). The resulting surface grafted pDVB microspheres were separated by vacuum filtration over a Millipore filter and washed successively three times with acetone, THF and ether. The polymeric particles were then dried in a vacuum. It should be noted that the modification of the surface of the pDVB microspheres, with either initiator or PPFS, shows an increase in surface

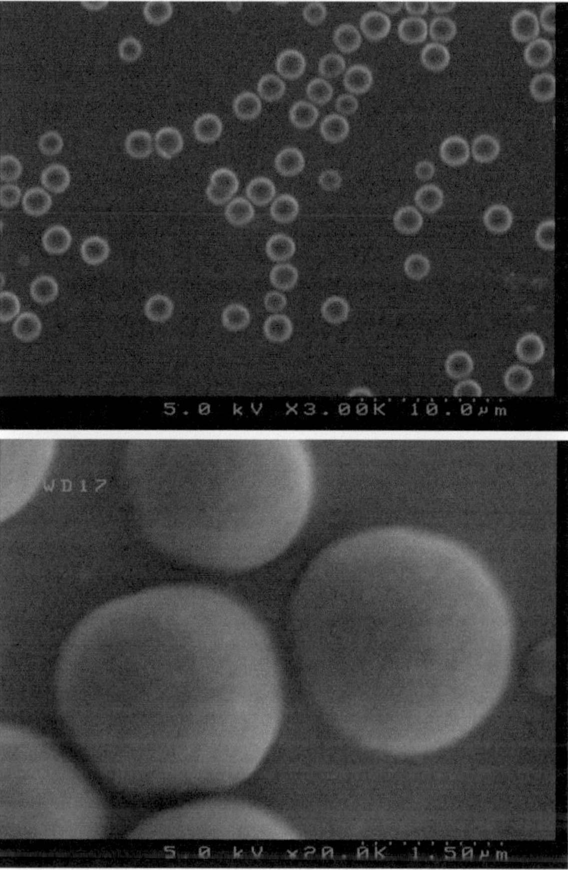

Fig. 4 SEM micrographs for the pDVB microspheres generated from precipitation-distillation polymerization

roughness according to the SEM images. In addition to the slight visible increase to surface roughness, the DLS and SEM images PPFS-pDVB microspheres exhibit an increase in particle diameter do to the surface modification. The diameter of PPFS-grafted microspheres was determined to be 1.7 μm using SEM, as opposed to 1.6 μm according to the DLS. This discrepancy may result from the hydrophobicity of PPFS. The partial collapsing of the hydrophobic PPFS brushes in water, which was used for DLS measurements, would account for this difference.

Finally, after synthesizing brushes from the microsphere surfaces and modifying them so as to make glycopolymer brushes, we investigated the protein binding of these materials so as to generate high binding affinity particles. As shown in Fig. 2, the thioglucosylated PPFS glycopolymer exhibits excellent binding affinity towards Con A. To show that this property can be

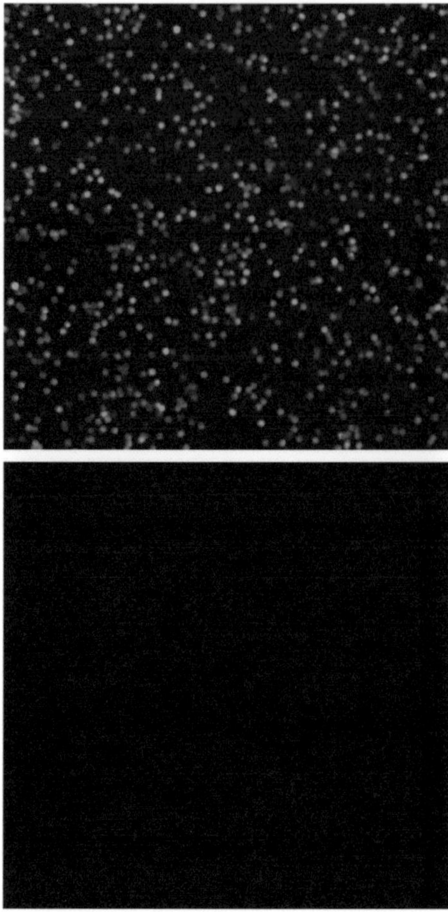

Fig. 5 Confocal microscopy images for FITC-Con A bound to Gluc-PPFS-pDVB microspheres (*left*) and PPFS-pDVB (*right*)

transferred to the modified microsphere particles, a fluorescently tagged Con A (FITC-Con A) was used to show the binding to the particles. Upon binding with the particle surface, the microspheres will fluoresce during confocal microscopy experiments due to the labelled FITC component. The confocal microscope image (Fig. 5) shows obvious florescence for glycopolymer modified pDVB microspheres and no florescence for PPFS modified pDVB microspheres, which was tested as a control. The binding to the surface is rather simple, with the only required caution being the rinsing of the microspheres with copious amounts of HEPES buffered water to ensure that any unbound FITC-Con A is removed. Unbound FITC-Con A will fluoresce in the confocal microscopy experiments as well, but this would be present in solution and show up as background scatter/fluorescence rather than specific fluorescence of the microsphere particles, as shown in the figure.

3.1 Synthesis of Polydivinylbenzene Microspheres

1. Apparatus consisting of a 250 mL round-bottom flask equipped with a magnetic stir bar connected to a condenser and a 100 mL receiving flask. The reaction flask is immersed in an oil bath atop a hot plate stirrer whilst the receiving flask is in an ice bath (*see* **Note 3**).
2. Dissolve AIBN (0.92 g, 0.56 mmol) and DVB80 (5 mL, 35.3 mmol) in 200 mL of acetonitrile and place in the reaction flask.
3. Heat the oil bath so that the solution begins boiling (roughly 85 °C) within 30 min of starting the reaction (*see* **Note 4**).
4. Stop the reaction after 100 mL of acetonitrile is distilled from the reaction system and collected in the receiving flask, which takes approximately 1.5 h.
5. Separate the resulting pDVB microspheres by vacuum filtration over a Millipore filter and wash three times each with THF, acetone, and ether, sequentially.
6. Dry the polymeric particles at 50 °C in a vacuum oven to afford 1.2 g of polymer microspheres (35 % yield) (*see* **Note 5**).

3.2 Synthesis of a Fluorinated Glycopolymer via Thiol Click Reaction to Poly(pentafluorostyrene)

1. In a 25 mL Schlenk tube, add copper(I) bromide (0.0602 g, 0.423), seal it with a rubber stopper, and then place it under vacuum for 1 h.
2. In a second Schlenk tube, add 10 mL of anisole and 6 mL of pentafluorostyrene (8.436 g, 43.5 mmol) and degas it via nitrogen bubbling for 30 min whilst sitting in an ice bath (*see* **Note 6**).
3. Transfer the liquid contents to the copper(I) bromide vessel with cannula needle whilst maintaining an inert environment.
4. Inject PMDETA (0.145 g, 0.836 mmol) via syringe and place the mixture in a 90 °C oil bath til a homogeneous solution is obtained, approximately 15 min.
5. Inject E2BriB (61.4 μL, 0.0816 g, 0.418 mmol) using a microsyringe to initiate the polymerization.
6. Pull aliquots of the reaction solution at timed intervals in order to determine the polymer conversion (via ^{1}H NMR) and molecular weight distributions (via SEC using DMAc as the eluent).
7. Thiol-halogen click reactions are performed using 0.07 g of an 18,000 g/mol PPFS sample (0.36 mmol of PFS units) and 0.1 g of 1-thio-β-D-glucose sodium salt (0.458 mmol) dissolved in 2 mL of dimethylformamide (DMF) with triethylamine as a catalyst (1.08 mmol, 0.15 mL) (*see* **Note 7**).
8. Allow the solution to react at room temperature for 6 h.

9. Precipitate the polymer in cold methanol.
10. The conversion and number average molecular weight is determined using ^{1}H and ^{19}F NMR (98 % conversion to glycopolymer).

3.3 Synthesis of a Fluorinated Glycopolymer Modified Microsphere via ATRP and Thiol Click Chemistry

1. Add polydivinylbenzene (pDVB) microspheres (0.5 g) and purified copper(I) bromide (60.2 mg, 0.423 mmol) in a 25 mL Schlenk tube and seal it and place under vacuum for 1 h.
2. Add anisole (10 mL) and pentafluorostyrene (6 mL, 43.5 mmol) in a second Schlenk tube and seal and degas it by bubbling nitrogen through the reaction solution whilst in an ice bath for 30 min (*see* **Note 6**).
3. Transfer the degassed solution over the copper(I) bromide vessel by a cannula.
4. To this mixture, add 0.175 mL of PMDETA (0.145 g, 0.836 mmol) and 61.4 μL of E2BriB (0.0816 g, 0.418 mmol) using a micro-syringe.
5. Heat the vessel to 90 °C and leave it to react for 16 h so as to result in PPFS-modified pDVB microspheres.
6. Separate the resulting surface modified pDVB microspheres by vacuum filtration over a Millipore filter and wash successively three times with THF, acetone, and ether followed by drying in a vacuum oven overnight. Collect and save the THF washings.
7. The THF washings contain "free" PPFS polymer. Precipitate the PPFS generated in solution during the surface initiated ATRP in cold methanol.
8. The filtered microspheres are characterized by ATR-FTIR, SEM, and dynamic light scattering (DLS).
9. Mix the polypentafluorostyrene-grafted pDVB microspheres (0.1 g) and 1-thio-β-D-glucose sodium salt (0.1 g, 0.46 mmol) in 2 mL of DMF with triethylamine as a catalyst (1.08 mmol, 0.15 mL).
10. Allow the mixture to react for 6 h at ambient temperature.
11. Collect the resulting semi-fluorinated glycopolymer-modified microspheres using a Millipore filter and wash with THF and acetone.
12. Analyze the microspheres using ATR-FTIR and DLS.

3.4 Protein Binding to the Fluorinated Glycopolymer

1. Dissolve Con A in 10 mM HEPES buffer solution (pH = 7.5) to afford a 30 μM solution.
2. Dissolve the semi-fluorinated glycopolymer samples in 10 mM HEPES buffer solution (pH = 7.5) to afford a solution containing 1.5 mM of glucose units.

3. Dilute the Con A buffered solution with an equal volume of 10 mM HEPES buffer solution (pH = 7.5) and monitor the transparency at 420 nm in a UV–Vis spectrometer to obtain an initial transmittance value.
4. Combine equal volumes, 1.5 mL, of the Con A buffered solution and the semi-fluorinated glycopolymer buffered solution and mix vigorously for 5 s before monitoring the transmittance at 420 nm to determine the extent of protein-glycopolymer cluster formation (*see* **Note 8**).

3.5 Protein Interaction Studies for Glycopolymer-Grafted pDVB Microspheres

1. Dissolve a solution of FITC-Con A (30 μM) in 10 mM HEPES buffer solution (pH 7.5).
2. In a separate vial, suspend glycopolymer-modified microspheres (Glyc-PPFS-pDVB, 20 mg) in 1 mL of 10 mM HEPES buffer solution (pH 7.5).
3. Combine equal parts of each solution, 0.5 mL and mix for 15 min.
4. Separate the resulting microspheres by vacuum filtration over a Millipore filter and wash three times with 1 mL of 10 mM HEPES buffer solution (pH 7.5).
5. Dry the microspheres were under vacuum and store in fridge prior to use. The same procedure is done using PPFS-grafted microspheres (PPFS-pDVB) so as to produce a control sample.
6. The two samples are both suspended in deionized water and measured by confocal microscope.

4 Notes

1. Removal of inhibitor present in all monomers shipped and received is critical to proper polymerization. The basic alumina in the column will turn yellow as the inhibitor reacts and binds to the basic alumina.
2. Cu(I)Br is white when in its proper oxidation state. As the material oxidizes, it will turn blue/green in color reducing its activity in the ATRP reaction. The copper bromide to be purified is made into a paste with 1 N sulphuric acid. The paste is mixed with 500 mL of sulfurous acid for 30 min before being rinsed with glacial acetic acid followed by anhydrous ether. The off-white precipitate is dried in a vacuum oven and stored in a desiccator when not in use.
3. To ensure solvent is distilled, rather than refluxed back into the reaction flask, the top of the reaction flask up to the first few centimeters of the condensing tube should be wrapped in

aluminium foil. This will ensure minimal heat loss during the reaction.

4. For good results, typically the oil bath is turned on and set to 85 °C prior to the addition of the reaction solution. This ensures that the system begins boiling and reacting within 30 min of being introduced to the oil bath.
5. This reaction setup is critical. Attempts to scale up the reaction results in broad particle size distributions as the distillation rate of the acetonitrile changes with respect to the polymerization rate. It is suggested that multiple reactions be performed rather than attempting to scale up the reaction size to obtain a larger quantity of particles.
6. The ice bath is to ensure that no monomer or solvent is volatilized during the degassing step.
7. Any PPFS molecular weight may be employed; however the amount of 1-thio-β-D-glucose sodium salt needs to be adjusted so that it is in molar excess to the amount of pentafluorostyrene units and *not* the molar amount of polymer.
8. The ConA will continually bind with the glycopolymer sample until saturation of the binding sites occurs. As the binding occurs, the solution will become more turbid and the transmittance will decrease with time. After a plateau in the transmittance, meaning full binding has occurred, the transmittance may increase. This is due to the bound polymer-protein system precipitating and settling to the bottom of the cuvette. These time points should be disregarded.

Acknowledgments

The authors are grateful for financial support from the Australian Research Council (ARC) in the form of a Discovery Grant (DP0877122).

References

1. Becer CR, Babiuch K, Pilz D, Hornig S, Heinze T, Gottschaldt M, Schubert US (2009) Clicking pentafluorostyrene copolymers: Synthesis, nanoprecipitation, and glycosylation. Macromolecules 42:2387–2394
2. Dere RT, Wang YX, Zhu XM (2008) A direct and stereospecific approach to the synthesis of α-glycosyl thiols. Org Biomol Chem 6:2061–2063
3. Dere RT, Zhu XM (2008) The first synthesis of a thioglycoside analogue of the immunostimulant KRN7000. Org Lett 10:461–4644
4. Zottola MA, Alonso R, Vite GD, Fraser-Reid B (1989) A practical, efficient large-scale synthesis of 1,6-anhydrohexopyranoses. J Org Chem 54:6123–6125
5. Aberg PM, Ernst B (1994) Facile preparation of 1,6-anhydrohexoses using solvent effects and a catalytic amount of a Lewis acid. Acta Chem Scand 48:228–233
6. Bai F, Yang XL, Huang WQ (2004) Synthesis of narrow or monodisperse poly(divinylbenzene) microspheres by distillation-precipitation polymerization. Macromolecules 37:9746–9752

7. Kolb HC, Finn MG, Sharpless KB (2001) Click chemistry: diverse chemical function from a few good reactions. Angew Chem Int Ed 40:2004–2021
8. Rosen BM, Lligadas G, Hahn C, Percec V (2009) Synthesis of dendritic macromolecules through divergent iterative thio-bromo 'click' chemistry and SET-LRP. J Polym Sci A Polym Chem 47:3940–3948
9. Samaroo D, Vinodu M, Chen X, Drain CM (2007) meso-Tetra(pentafluorophenyl)porphyrin as an efficient platform for combinatorial synthesis and the selection of new photodynamic therapeutics using a cancer cell line. J Comb Chem 9:998–1011
10. Tan BH, Gudipati CS, Hussain H, He C, Liu Y, Davis TP (2009) Synthesis and self-assembly of pH-responsive amphiphilic poly(dimethylaminoethyl methacrylate)-block-poly(pentafluorostyrene) block copolymer in aqueous solution. Macro Rap Comm 30:1002–1008
11. Granville AM, Boyes SG, Akgun B, Foster MD, Brittain WJ (2004) Synthesis and characterization of stimuli-responsive semifluorinated polymer brushes prepared by atom transfer radical polymerization. Macromolecules 37:2790–2796
12. Lin SS, Levitan IB (1991) Concanavalin A: a tool to investigate neuronal plasticity. Trends Neurosci 14:273–277
13. Chen GJ, Tao L, Mantovani G, Geng J, Nystrom D, Haddleton DM (2007) A modular click approach to glycosylated polymeric beads: design, synthesis and preliminary lectin recognition studies. Macromolecules 40:7513–7520
14. Jain S, Gupta MN (2005) An integrated process for separation of major and minor proteins from goat serum. Appl Biochem Biotechnol 125:53–62
15. Pfaff A, Barner L, Mueller AHE, Granville AM (2011) Surface modification of polymeric microspheres using glycopolymers for biorecognition. Euro Polym J 47:805–815

Chapter 11

Glycopolymer-Grafted Polymer Particles for Lectin Recognition

Michinari Kohri, Tatsuo Taniguchi, and Keiki Kishikawa

Abstract

Glycopolymers bearing carbohydrates have an advantage in protein recognition that is attributable to the multivalent effect (cluster effect) of side-chain carbohydrates. A variety of surface-modified polymer particles have been prepared concurrently with the development of new synthetic technology. Here we describe a synthetic method of glycopolymer-grafted polymer particles by surface-initiated living radical polymerization, i.e., atom-transfer radical polymerization (ATRP) and photoiniferter polymerization, for specific lectin recognition.

Key words Glycopolymers, Core–shell, Polymer particles, Emulsifier-free emulsion polymerization, Atom-transfer radical polymerization, Photoiniferter polymerization, Lectin

1 Introduction

Carbohydrates play important roles for biomolecular recognition such as cell–cell interaction, signal transmission, and inflammation [1, 2]. Glycopolymer-modified materials have attracted much attention in the biomedical field, since glycopolymers containing carbohydrates have an advantage in protein recognition attributable to the multivalent effect (so called cluster effect) of side-chain carbohydrates [3]. Glycopolymer-modified gold, silica, quantum dots, and magnetic particles have been prepared for lectins (proteins with carbohydrate-binding domains) recognition materials [4–7]. Many researchers have also reported glycopolymer-bearing polymer particles using self-assemblies of amphiphilic di-block copolymers containing a glycopolymer [8] or surface-initiated free radical polymerizations on polymer particles [9]. Recent progress in living radical polymerization techniques, such as atom-transfer radical polymerization (ATRP), reversible addition-fragmentation chain-transfer (RAFT) polymerization, nitroxidemediated radical polymerization, and iniferter polymerization, opens new routes for

Xue-Long Sun (ed.), *Macro-Glycoligands: Methods and Protocols*, Methods in Molecular Biology, vol. 1367, DOI 10.1007/978-1-4939-3130-9_11,

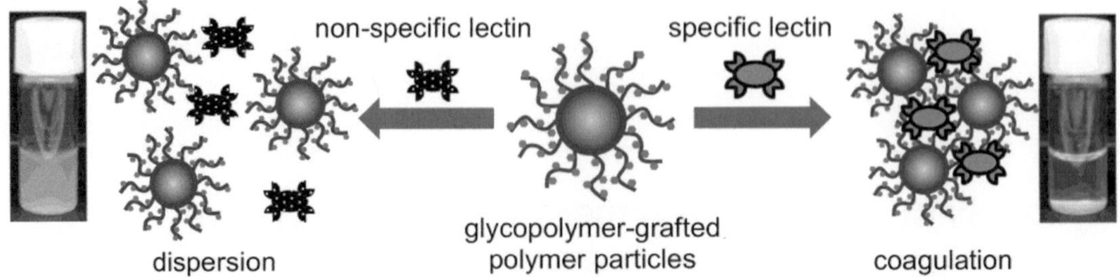

Fig. 1 Schematic representation of the coagulation of glycopolymer-grafted polymer particles in the presence of lectins

the synthesis of polymers with predefined molecular weights and narrow polydispersities [10].

The outline of the present chapter is as follows; we report the synthesis of glycopolymer-grafted polymer particles for lectin recognition using two types of surface-initiated living radical polymerization: the first is ATRP, and the other is photoiniferter polymerization (Fig. 1). In accordance with the first one, ATRP initiating groups are introduced onto the surface of polymer particles by emulsifier-free emulsion polymerization of styrene (St) and 2-chloropropyloxyethyl methacrylate (CPEM). Glycomonomers containing glucose or lactose residues are then polymerized onto the core particles using surface-initiated ATRP. In the second method, ATRP initiating groups onto polymer particles are replaced for the photoiniferter group using a photosensitive iniferter, i.e., sodium *N*,*N*-diethyldithiocarbamate (NaDC). The graft polymerization of a glycomonomer is initiated by UV irradiation (400 W, $\lambda = 365$ nm). Moreover, particles bearing di-block copolymer shells, using glycomonomer and polyethylene glycol monomethacrylate as monomers, are prepared. Both synthetic routes are considered in detail below. Also we describe the binding abilities of glycopolymer-grafted polymer particles against two types of lectins, peanut agglutinin (PNA) and concanavalin A (Con A). PNA includes four identical subunits, and each subunit can recognize a galactose residue [11]. Similarly, Con A has four identical subunits, but each subunit specifically binds glucose and mannose residues [12].

2 Materials

2.1 Chemicals

1. Styrene (St) (Kanto Chemical Co., Inc.) is dried over calcium hydride and distilled under reduced pressure.
2. Potassium persulfate (KPS).
3. 2-Chloropropyloxyethyl methacrylate (CPEM) (*see* **Note 1**).
4. 4-Vinylbenzenesulfonamidoethyl 1-thio-β-D-glucopyranoside (glycomonomer **1**) (*see* **Note 2**).

5. 4-Vinylbenzenesulfonamidoethyl 1-thio-β-D-lactoside (glycomonomer **2**) (*see* **Note 3**).
6. Copper (II) dichloride dihydrate ($CuCl_2 \cdot 2H_2O$).
7. Tris[2–(dimethylamino)ethyl]amine (Me_6TREN) (*see* **Note 4**).
8. Methanol.
9. Water used in all experiments is distilled and deionized by a Millipore Simplicity.
10. L-Ascorbic acid.
11. *N,N*-diethyldithiocarbamate (NaDC).
12. Anthrone.
13. Sulfuric acid.
14. Polyethylene glycol (PEG) monomer **3** (Blemmer PE-200, $n = 4, 5$, NOF Co).
15. Phosphate-buffered saline.
16. 2-Hydroxyethyl methacrylate.
17. 2-Chloropropionyl chloride.
18. Tetrahydrofuran (THF).
19. Diethyl ether.
20. Sodium hydrogen carbonate ($NaHCO_3$).
21. Sodium sulfate.
22. Ethyl acetate.
23. Hexane.
24. Per-*O*-acetylated glucopyranosyl bromide.
25. Thiourea.
26. Acetonitrile (CH_3CN).
27. Triethylamine.
28. 2-Bromoethylamine hydrobromide.
29. Dichloromethane (CH_2Cl_2).
30. Triethylamine.
31. 4-Vinylbenzenesulfonyl chloride.
32. Per-*O*-acetylated lactosyl bromide.
33. Formaldehyde (37 % (w/w)).
34. Formic acid (98 % (w/w)).
35. Tris (2-aminoethyl) amine (TREN).

2.2 Lectins

1. Peanut agglutinin (PNA).
2. Concanavalin A (Con A).

2.3 Instruments

1. Tubing pump system (Master Flex L/S; Cole-Parmer Instrument Co.).
2. Filter membrane (50 nm pore size; Spectrum Laboratories Inc.).

3 Methods

Two methods of preparation of glycopolymer-grafted polymer particles are described. In the first method adopted from literature [13], surface-initiated ATRP are used for preparation of glycopolymer shell layer onto polymer particles. Firstly, ATRP-initiating groups-bearing core particles are prepared using emulsifier-free emulsion polymerization of St with CPEM: P(St-CPEM) core particles (Fig. 2). The total ATRP-initiator concentration from ^{1}H nuclear magnetic resonance (NMR) measurements (freeze-dried sample dissolved in $CDCl_3$) is 4.80×10^{-4} mol/g, which is comparable to the amount of CPEM polymerized in the reaction (5 mol%). The surface initiator concentration is calculated from a conductometric titration. Next, surface-initiated ATRP of glycomonomers **1** and **2** are performed. The volume-average diameters of **GP1(15)** and **GP2(25)** particles in water, as measured using dynamic light scattering (DLS), are 400 nm and 420 nm, respectively, which are larger than the 370 nm of the core P(St-CPEM), indicating that **GP1(15)** and **GP2(25)** have hydrated glycopolymer layers of 15 nm and 25 nm thickness, respectively. From scanning electron microscope (SEM) images, dried **GP1(15)** and **GP2(25)** particles apparently maintain monodispersity after grafting glycopolymer.

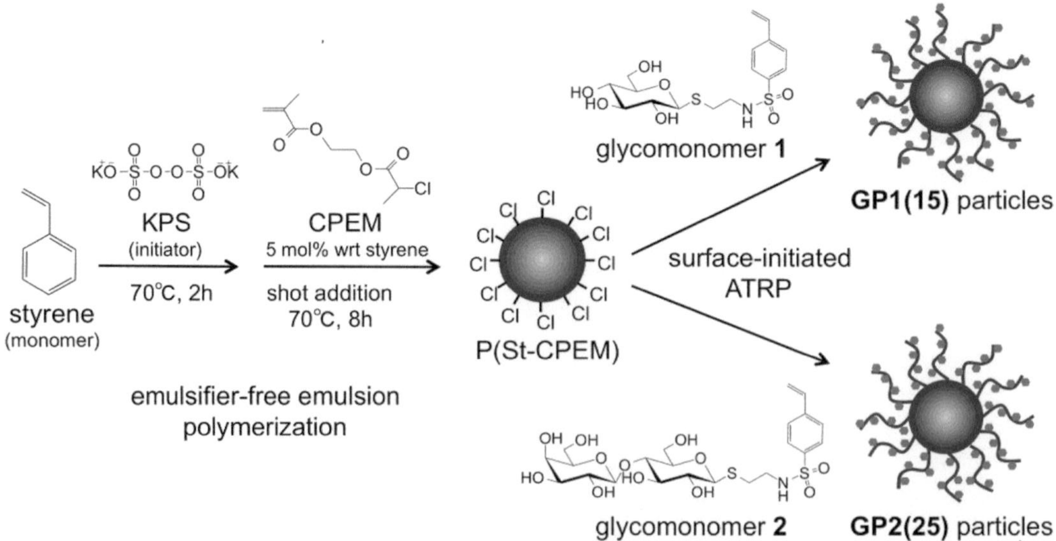

Fig. 2 Preparation of glycopolymer-grafted polymer particles by surface-initiated ATRP

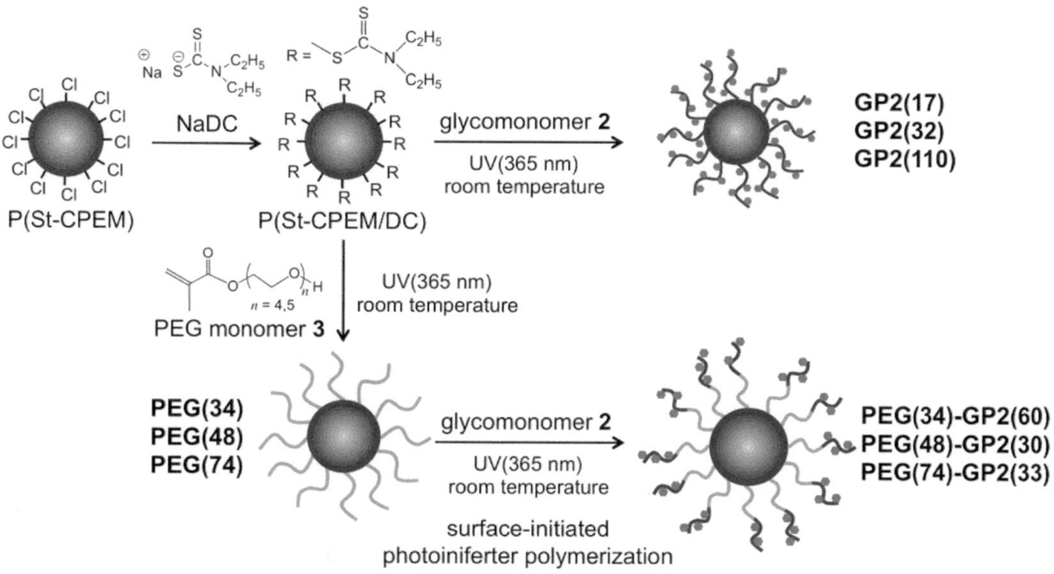

Fig. 3 Preparation of glycopolymer-grafted polymer particles by surface-initiated photoiniferter polymerization

In the second method adopted from literature [14], surface-initiated photoiniferter polymerization is used for preparation of glycopolymer-grafted particles. NaDC is immobilized on the P(St-CPEM) particles at room temperature, giving rise to the photoiniferter-attached particles: P(St-CPEM/DC) particles (Fig. 3). P(St-CPEM/DC) particles are dispersed in an aqueous solution of the glycomonomer **2**, and the dilute solution is exposed to UV light (400 W, $\lambda = 365$ nm). The glycopolymer shell thickness gradually increases with increasing reaction time or glycomonomer concentration. The **GP2(17)**, **GP2(32)**, and **GP2(110)** particles, have hydrated glycopolymer layer thicknesses of 17 nm, 32 nm, and 110 nm, respectively, are hence chosen for the lectin-binding assay. The PEG polymer-grafted particles are prepared similarly to the glycopolymer-grafted particles. The **PEG(34)**, **PEG(48)**, and **PEG(74)** particles with different PEG polymer shell thicknesses of 34 nm, 48 nm, and 74 nm, respectively, are used for the preparations of particles bearing a di-block copolymer shell. By controlling the reaction conditions, the **PEG(34)-GP2(60)**, **PEG(48)-GP2(30)**, and **PEG(74)-GP2(33)** particles have hydrated glycopolymer layer thicknesses of 60 nm, 30 nm, and 33 nm, respectively, are prepared. The amount of lactose conjugated onto the P(St-CPEM/DC) particles is quantified by the anthrone-sulfuric acid method (*see* **Note 5**). These analytical data for the particles obtained are shown in Table 1. The lactose amount gradually increase from 16.3 up to 57.5 μg/mg (μg of carbohydrate moieties per mg of particles) as the glycopolymer shell thickness increase (entries 1–3). The amount of lactose on the

Table 1
The amounts of carbohydrate residues conjugated to the particles measured by anthrone-sulfuric acid method

Entry	Sample	Carbohydrate residues per particle (μg/mg)
1	GP2(17)	16.3
2	GP2(32)	30.3
3	GP2(110)	57.5
4	PEG(34)-GP2(60)	36.2
5	PEG(48)-GP2(30)	22.4
6	PEG(74)-GP2(33)	28.5

glycoparticles is dependent on the shell thickness of the core–shell particles. For the di-block copolymer-grafted particles, the conjugated lactose amount also increases as the glycopolymer shell layers increased (entries 4–6).

Finally, we describe the lectin-binding assay of the particles obtained. The specific lectin-binding abilities of the particles are analyzed using a turbidity assay. After addition of PNA or Con A solution to particles, the mixture is left to stand for 20 h to allow the particles to coagulate and settle to the bottom. Then, the transmittance of the supernatant at 600 nm is measured to evaluate the lectin-binding ability of the latex. To investigate the effect of carbohydrate type on specific lectin recognition, **GP1(15)** and **GP2(25)** particles are used. The effect of glycopolymer shell thickness on lectin recognition, **GP2(17)**, **GP2(32)**, and **GP2(110)** particles are also compared. The transmittance of the detected in turbidity assay is plotted against lectin concentration and reaction time (Fig. 4). Results indicate that short glycopolymer-grafted particles allow lectins to act as a cross-linker between particles and are precipitated efficiently (Fig. 5). In contrast, the coagulation of long glycopolymer-grafted particles is inhibited because lectins mainly interacte with each surface of particles (Fig. 5). The introduction of PEG moieties in shell layer are also investigated using di-block copolymer-grafted particles, i.e., **PEG(34)-GP2(60)**, **PEG(48)-GP2(30)**, and **PEG(74)-GP2(33)** particles.

To improve the assay sensitivity for lectin recognition of **GP2(17)** and **PEG(48)-GP2(30)** particles, DLS measurements are performed because DLS is a powerful tool that can detect small changes in the aggregation of particles due to the presence of the lectins [15]. Size distributions of the particles in the presence of lectins are measured by DLS. The introduction of PEG moieties at the inner positions of the shell layers are shown to inhibit the adhesion of proteins,

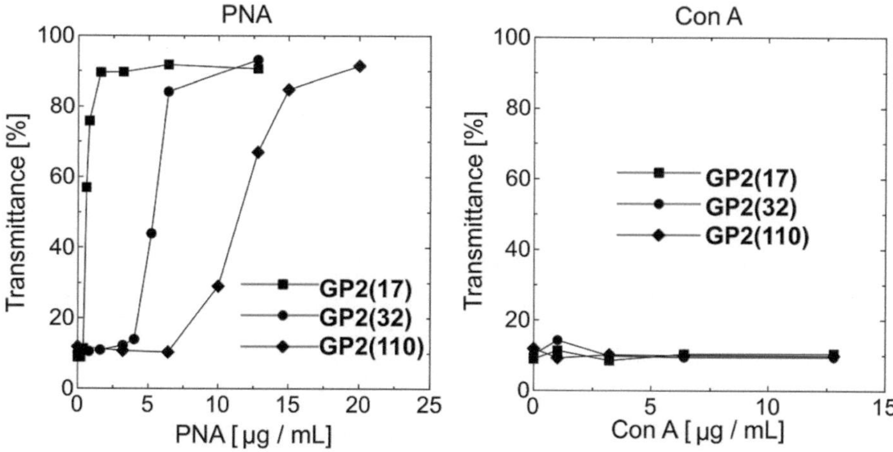

Fig. 4 Transmittance, at 600 nm, of **GP2(17)**, **GP2(32)**, and **GP2(110)** particles-lectin mixtures as a function of the lectin concentration. The lectin solution was added to the polymer latex (0.066 wt%) and allowed to stand for 20 h at room temperature

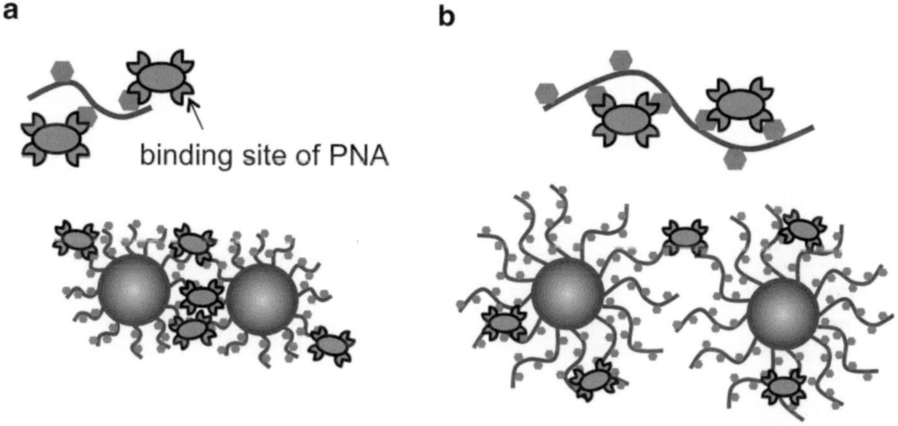

Fig. 5 Binding models of (**a**) short glycopolymer-grafted particles-lectin and (**b**) long glycopolymer-grafted particles-lectin

indicating more precise detection of lectin recognition without non-specific adsorption of the lectins is achieved with PEG-containing di-block copolymer-grafted particles.

3.1 Preparation of Glycopolymer-Grafted Polymer Particles by Surface-Initiated ATRP

3.1.1 Preparation of Polystyrene Particles Bearing ATRP-Initiating Groups (P(St-CPEM))

1. Add St (3.52 g, 33.8 mmol), KPS (0.18 g, 0.65 mmol), and deionized water (100 mL) in a three-necked flask, and deoxygenate by purging with argon for 15 min.
2. Initiate the polymerization by heating to 70 °C in a water bath.
3. After stirring for 2 h, add CPEM (0.46 g, 2.1 mmol). Then continue the polymerization for another 6 h.
4. Cool the reaction solution in an ice bath and filter it to remove any aggregates that are present.

5. Dialyze the filtrate solution with a hollow fiber dialyzer using a filter membrane and a tubing pump system to remove unreacted monomers and initiator and afford P(St-CPEM) after freeze-dry.
6. The surface initiator concentration is calculated from a conductometric titration (*see* **Note 6**).

*3.1.2 Preparation of **GP1(15)** and **GP2(25)** Particles by Surface-Initiated ATRP*

1. Add glycomonomer (**1** or **2**), $CuCl_2 \cdot 2H_2O$, Me_6TREN, P(St-CPEM) latex (4.9 wt%), and the solvent (10 mL) in a round-bottom flask. The solvents used are, respectively, a 1:4 methanol/water mixture and water for preparation of **GP1** and **GP2** (*see* **Note** 7).
2. Deoxygenate the mixture by purging with argon for 15 min, with subsequent placement in a water bath at 30 °C.
3. To the solution above, add an argon-purged aqueous solution of L-ascorbic acid.
4. After stirring for 24 h, stop the polymerization by purging with oxygen.
5. Collect the particles and purify repeatedly by centrifugation (17,730 × *g*, 15 min) and redispersion.

3.2 Preparation of Glycopolymer-Grafted Polymer Particles by Surface-Initiated Photoiniferter Polymerization

3.2.1 Preparation of the Polystyrene Particles Bearing Photoiniferter-Initiating Groups (P(St-CPEM/DC))

1. To the P(St-CPEM) particles (1.00 g) and deionized water (52 mL), add a NaDC solution (0.51 g, 2.3 mmol in a small amount of water) dropwise at 0 °C.
2. After stirring for 24 h at room temperature, collect the obtained P(St-CPEM/DC) latex and purify repeatedly by centrifugation (17,730 × *g*, 15 min) and redispersion.

*3.2.2 Preparation of **GP2(17)**, **GP2(32)**, and **GP2(110)** Particles by Surface-Initiated Photoiniferter Polymerization*

1. Add glycomonomer **2** (0.114–0.568 g, 0.2–1.0 mmol), P(St-CPEM/DC) particles (0.05 g), and deionized water (20 mL) in a round-bottom flask (*see* **Note 8**).
2. Deoxygenate the mixture by purging with argon for 10 min, and then place the flask in a water bath at room temperature and exposed to UV light (400 W, $\lambda = 365$ nm).
3. After 5–120 min, collect the particles and purify repeatedly by centrifugation (17,730 × g, 15 min) and redispersion.
4. Conjugated lactose amounts on the particles are measured by the anthrone-sulfuric acid method (*see* **Note 5**).

*3.2.3 Preparation of **PEG(34)-GP2(60)**, **PEG(48)-GP2(30)**, and **PEG(74)-GP2(33)** Particles by Surface-Initiated Photoiniferter Polymerization*

1. Add PEG monomer **3** (0.994 g, 3.5 mmol), P(St-CPEM/DC) particles (0.05 g), and deionized water (20 mL) in a round-bottom flask.
2. Deoxygenate the mixture by purging with argon for 10 min, and then place the flask in a water bath at room temperature and exposed to UV light (400 W, $\lambda = 365$ nm).
3. After 5–120 min, collect the particles and purify repeatedly by centrifugation (17,730 × *g*, 15 min) and redispersion.
4. Add glycomonomer **2** (0.114–0.227 g, 0.2–0.4 mmol), PEG polymer-grafted particles (0.025 g), and deionized water (10 mL) in a round-bottom flask.
5. Deoxygenate the mixture by purging with argon for 10 min, and then place the flask in a water bath at room temperature and exposed to UV light (400 W, $\lambda = 365$ nm).
6. After 60 or 120 min, collect the particles and purify repeatedly by centrifugation (17,730 × *g*, 15 min) and redispersion.
7. Conjugated lactose amounts on the particles are measured by the anthrone-sulfuric acid method (*see* **Note 4**).

3.3 Lectin-Binding Tests

3.3.1 Turbidity Assay

1. Add a 10 mM lectin solution in 10 mM phosphate-buffered saline (pH 7.4) to latex (0.066 wt%) (*see* **Note 9**).
2. Leave the mixture to stand for 20 h to allow particles to coagulate and settle to the bottom.
3. Measure the transmittance of the supernatant at 600 nm to evaluate the lectin binding ability of the latex.

3.3.2 Dynamic Light Scattering Measurements

1. Add a lectin solution in 10 mM phosphate-buffered saline (pH 7.4) (0.70 μg/mL) to latex solution (8.0×10^{-4} wt%) (*see* **Note 9**).
2. After 90 min incubation, measure the size distributions of samples by DLS.

4 Notes

1. 2-Hydroxyethyl methacrylate (17.6 g, 135 mmol) and pyridine (21.8 g, 276 mmol) in THF (150 mL) is added 2-chloropropionyl chloride (25.0 g, 198 mmol) in THF in a dropwise manner at 0 °C. After stirring for 5 h at room temperature, the reaction mixture is diluted with diethyl ether and wash with saturated aqueous $NaHCO_3$. The organic layer is dried over sodium sulfate. After filtration, the solvent is evaporated to dryness. The residue is purified using silica gel chromatography with ethyl acetate/hexane as a mobile phase.
2. The mixture of commercially available per-*O*-acetylated glucopyranosyl bromide (10.86 g, 26.4 mmol) and thiourea (2.77 g,

36.5 mmol) in CH_3CN (40 mL) is stirred at 60 °C for 2.5 h. After cooling to room temperature, triethylamine (13.6 mL, 97.4 mmol) and 2-bromoethylamine hydrobromide (7.47 g, 36.5 mmol) are added reaction mixtures. After stirring for 2 h at room temperature, the reaction mixture is diluted with CH_2Cl_2 and wash with saturated aqueous $NaHCO_3$. The organic layer is dried over sodium sulfate. After filtration, the solvent is evaporated to produce aminoglycoside. Then, aminoglycoside (23.6 mmol) and triethylamine (13.5 mL, 96.7 mmol) in CH_2Cl_2 (50 mL) is added 4-vinylbenzenesulfonyl chloride (7.17 g, 35.4 mmol) in CH_2Cl_2 (30 mL) in a dropwise manner at 0 °C. After stirring for 24 h at room temperature, the reaction mixture is diluted with CH_2Cl_2 and wash with saturated aqueous $NaHCO_3$. The organic layer is dried over sodium sulfate. After filtration, the solvent is evaporated to dryness. The residue is purified using silica gel chromatography with ethyl acetate/hexane as a mobile phase.

3. Glycomonomer **2** is synthesized similarly to glycomonomer **1**. Per-*O*-acetylated lactosyl bromide is used as a starting material.
4. A mixture of formaldehyde (37 % (w/w), 60.5 g, 1.29 mol) and formic acid (98 % (w/w), 36.0 g, 0.432 mol) is stirred at 0 °C. After 1 h, a solution of TREN (12.3 g, 0.084 mol) and deionized water (17.5 g) is added in a dropwise manner. The mixture is gently refluxed overnight at 95 °C. After cooling to room temperature, the volatile fractions are removed by rotary evaporation. The residue is treated with a saturated sodium hydroxide aqueous solution until pH > 10, producing an oil layer, which is extracted into CH_2Cl_2. The organic layer is dried over sodium sulfate. After filtration, the solvent is evaporated to produce yellow oil.
5. Dilute hydrochloric acid (final concentration: 1.2–1.4 M) is added to the latex, and then the mixture is stirred for 24 h at 80 °C. The particles are separated by centrifugation (17,730 × *g*, 45 min), and an anthrone (0.2 %) solution in aqueous sulfuric acid (1.2 mL, concentrated sulfuric acid:water = 5:2) is added to the supernatant (0.2 mL). The mixture is vigorously stirred for 10 min at 100 °C, and then the transmittance of the reaction mixture at 620 nm is measured. Lactose is used as the standard sample.
6. The latex is stirred with 1 M sodium hydroxide for 36 h at room temperature. The suspension is multiply centrifuged. Then the pellet is resuspended in deionized water. The centrifugation and redispersion are continued until pH 7.0. Conductometric titration gives a total negative charge on the surface, representing the sum of the sulfate surface charge from

KPS and the carboxyl groups produced from hydrolysis of the ATRP-initiator (CPEM). The surface charge of the latex is determined from the conductometric titration of unhydrolyzed latex; the difference between the two values indicates the surface concentration of the initiator.

7. The solubility of **1** in water is rather low as a result of the presence of the hydrophobic styrene moiety.
8. An addition of glycomonomer above 50 mM results in coarse aggregation of the particles and gelation, indicating that an excess addition of **2** causes an increase in viscosity.
9. Freshly prepare immediately before use.

Acknowledgments

This work was partially supported by a Grant-in-Aid for Scientific Research from the Ministry of Education, Culture, Sports, Science and Technology of Japan (Nos. 20850005 and 23750119).

References

1. Wulff G, Schmid J, Venhoff T (1996) The synthesis of polymerizable vinyl sugars. Macromol Chem Phys 12:259–274
2. Dwek RA (1996) Glycobiology: toward understanding the function of sugars. Chem Rev 96:683–720
3. Ladmiral V, Melia E, Haddleton DM (2004) Synthetic glycopolymers: an overview. Eur Polym J 40:431–449
4. Boyer C, Bousquet A, Rondolo J et al (2010) Glycopolymer decoration of gold nanoparticles using a LbL approach. Macromolecules 43:3775–3784
5. Guo TY, Liu P, Zhu JW et al (2006) Well-defined lactose-containing polymer grafted onto silica particles. Biomacromolecules 7: 1196–1202
6. Jiang X, Ahmed M, Deng Z et al (2009) Biotinylated glyco-functionalized quantum dots: synthesis, characterization, and cytotoxicity studies. Bioconjug Chem 20:994–1001
7. Sun XL, Cui W, Haller C et al (2004) Site-specific multivalent carbohydrate labeling of quantum dots and magnetic beads. Chembiochem 5:1593–1596
8. Serizawa T, Yasunaga S, Akashi M (2001) Synthesis and lectin recognition of polystyrene core-glycopolymer corona nanospheres. Biomacromolecules 2:469–475
9. Revilla J, Elaïssari A, Carriere P et al (1996) Adsorption of bovine serum albumin onto polystyrene latex particles bearing saccharidic moieties. J Colloid Interface Sci 180:405–412
10. Braunecker WA, Matyjaszewski K (2007) Controlled/living radical polymerization: features, developments, and perspectives. Prog Polym Sci 32:93–146
11. Miller RL (1978) Purification of peanut (*Arachis hypogaea*) agglutinin isolectins by chromatofocusing. Anal Biochem 190:693–698
12. Renauer D, Oesch F, Kinkel J et al (1985) Fractionation of membrane proteins on immobilized lectins by high-performance liquid affinity chromatography. Anal Biochem 151: 424–427
13. Kohri M, Sato M, Abo F et al (2011) Preparation and lectin binding specificity of polystyrene particles grafted with glycopolymers bearing S-linked carbohydrates. Eur Polym J 47:2351–2360
14. Kohri M, Abo F, Miki S et al (2013) Effects of graft shell thickness and compositions on lectin recognition of glycoparticles. J Colloid Sci Biotechnol 2:45–52
15. Deng Z, Li S, Jiang X et al (2009) Well-defined galactose-containing multi-functional copolymers and glyconanoparticles for biomolecular recognition processes. Macromolecules 42: 6393–6405

Chapter 12

Synthesis of Non-spherical Glycopolymer-Decorated Nanoparticles: Combing Thiol-ene with Catecholic Chemistry

Xiao Li, Weidong Zhang, and Gaojian Chen

Abstract

Glycopolymers with carbohydrate side chains are currently being applied in many fields, with much potential for disease treatment. The shape of glycopolymer-bearing nanoparticles has obvious effects on the nanoparticle-cell interaction and is therefore important for the applications of glycopolymers in biological systems. Here a synthetic approach to prepare non-spherical glycopolymer-coated iron oxide nanoparticles is provided, by combing the convenience of inorganic shape control, catecholic chemistry, and thiol-ene reaction.

Key words Glycopolymer, Iron oxide, Nanoparticles, Dopamine

1 Introduction

Carbohydrates play complex roles in vivo, as an important ligand for interacting with receptors on cell surfaces. The specific affinity between carbohydrates and proteins can be greatly enhanced through the cluster glycoside effect. To better understand these biological interactions, and to broaden applications, glyconanoparticles, or more specifically glycopolymer-decorated nanoparticles, which are able to interact with lectin as multivalent ligand, have raised much attention. Efficient and versatile techniques to synthesize well-defined glyconanoparticles are needed, especially for non-spherical ones, considering the importance of nanoparticle shapes on cellular uptake and applications beyond [1, 2].

Versatile catecholic chemistry has been extensively used to create virtually all types of material surfaces, regardless of their chemical functionality or surface energy [3]. Non-spherical iron oxide nanoparticles, which can be synthesized easily and precisely with desired shape are modified to incorporate desired functionalities (vinyl groups in this example) by catecholic chemistry, and

Xue-Long Sun (ed.), *Macro-Glycoligands: Methods and Protocols*, Methods in Molecular Biology, vol. 1367, DOI 10.1007/978-1-4939-3130-9_12, © Springer Science+Business Media New York 2016

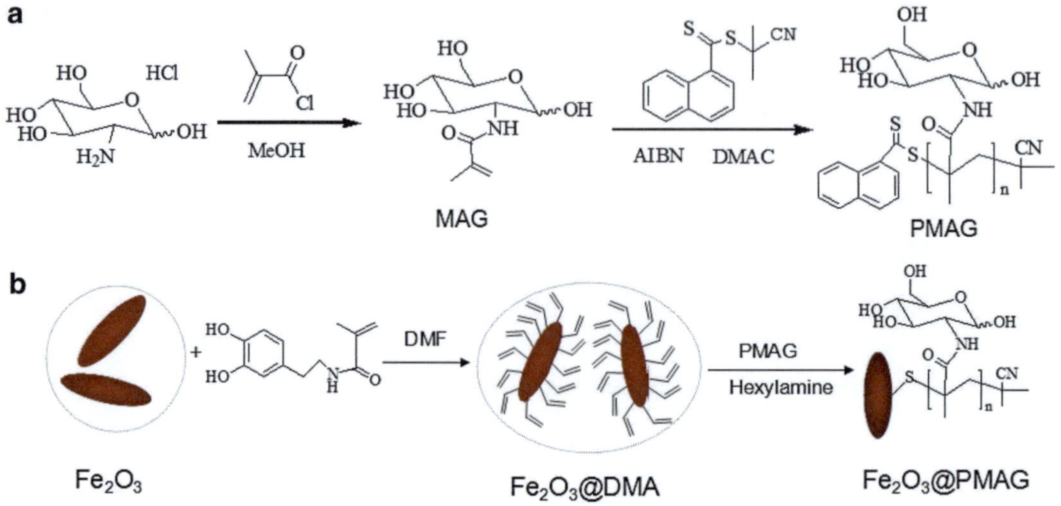

Fig. 1 (**a**) Preparation of PMAG via RAFT polymerization, (**b**) synthesis of PMAG decorated iron oxide nanoparticles [4]. Reproduced by permission of The Royal Society of Chemistry (http://pubs.rsc.org/)

then well-defined glycopolymers were synthesized by reversible addition-fragmentation chain transfer (RAFT) polymerization and conjugated with the pretreated iron oxide nanoparticles via thiol-ene reaction [4] (Fig. 1). The synthesized non-spherical glycopolymer-coated iron oxide nanoparticles could be used for further applications. In this chapter, spindle Fe_2O_3 nanoparticles are synthesized and used as an example. The approach could also be used for magnetic Fe_3O_4 nanoparticles or other nanoparticles that dopamine-based molecules would adhere to.

As shown in Fig. 1, glycopolymer was first synthesized by RAFT polymerization of 2-(methacrylamido)glucopyranose (MAG) [5], using 2-cyanoprop-2-yl-α-dithionaphthalate (CPDN) as the chain transfer agent (CTA) and 2,2′-azobis(isobutyronitrile) (AIBN) as the initiator [6]. Glycopolymers with various molecular weights could be obtained by controlling the polymerization time and initial ratios between monomer and CTA. Representative glycopolymers obtained with different monomer/CTA ratios and polymerization times are given in Table 1.

A biomimetic coating strategy was then chosen to modify the iron oxide surface by introducing vinyl groups through catecholic chemistry. Dopamine methacrylamide (DMA), a dopamine derivative with vinyl functionality was used for iron oxide surface modified. The vinyl groups introduced by DMA would then react with thiol-terminal glycopolymers via thiol click chemistry, therefore anchoring the polymer on the surface. Different analytical techniques including scanning electron microscopy (SEM), powder X-ray diffraction (XRD), thermogravimetric analysis (TGA), and Fourier transform infrared spectroscopy (FTIR) were used to evidence and to quantify glycopolymer coated on iron oxide.

Table 1
Glycopolymers obtained by polymerization of MAG with CPDN at 70 °C in DMAc with various polymerization time and the monomer/CTA ratios

Polymers	1	2	3	4	5
M/CTA	50.6/1	50.6/1	50.6/1	38/1	25.3/1
Time (h)	6	16	24	24	24
Conversion (%)	34	58	67	66	74
$M_{n,GPC}$ (g/mol)	7 100	14 700	20 000	10 200	5800
PDI	1.21	1.18	1.13	1.15	1.25

To test the affinity and uptake behaviors of the glycopolymer-coated iron oxide nanoparticles toward cancer cells for potential bio-labeling or imaging applications, HeLa cells could be chosen as a model.

2 Materials and Equipments

2.1 Chemicals

1. 2,2′-Azobis(isobutyronitrile) (AIBN) is recrystallized three times from ethanol.
2. 2-Cyanoprop-2-yl-α-dithionaphthalate (CPDN) is synthesized as described in ref. 7.
3. Glucosamine hydrochloride.
4. Fe_3O_4 nanoparticles.
5. Dopamine methacrylamide (DMA) are synthesized and characterized according to a previously reported method [8].
6. HeLa cells are cultured in Dulbecco's modified Eagle's medium (DMEM) at 37 °C and equilibrated in 5 % CO_2 and air.

2.2 Equipment

1. Microplate reader, SpectraMax M5, Molecular Devices, Sunnyvale, CA, USA.
2. Inductively coupled plasma-optical emission spectrometer (ICP-OES), Varian 710-ES, USA.
3. Freeze-dryer, LABCONCO FreeZone, 2.5 L, USA.
4. Centrifugal machine, Xiangyi TGL-20 M, China.
5. Sonifier, Scientz SB-3200DT, China.

3 Methods

3.1 Synthesis of 2-(methacrylamido) glucopyranose (MAG)

1. Add glucosamine hydrochloride (10.0 g, 4.64×10^{-2} mol) and potassium carbonate (6.41 g, 4.64×10^{-2} mol) in 250 mL of methanol in a 500 mL single-neck round-bottom flask and vigorously stir to dissolve it.

2. Cool the flask to −10 °C using an acetone/ice bath.
3. Add methacryloyl chloride (4.36 g, 4.17×10^{-2} mol) dropwise into the mixture with vigorous stirring (*see* **Note 1**).
4. Stir the mixture at −10 °C for 30 min and leave it to react for 3 h at room temperature.
5. After completion of the reaction, remove the precipitated salt via Buchner filtration and wash with methanol.
6. Collect the combined filtrate and concentrate it via rotary evaporation to afford an off-white slurry (<100 mL).
7. Load the slurry onto a silica gel column chromatograph for purification with dichloromethane-methanol (4:1 ratio) as the eluent.
8. Concentrate the collected colorless eluent portion by a rotary evaporator, until the solvent completely removed (*see* **Note 2**).
9. Yield of the product ~45 %; fine white powder (*see* **Note 3**).

3.2 Preparation of Poly(2-(methacrylamido) glucopyranose) (PMAG)

1. Add MAG (0.42 g, 1.7×10^{-3} mol), CPDN (9.13 mg, 3.37×10^{-5} mol), AIBN (0.77 mg, 4.71×10^{-6} mol) into a 5 mL ampule containing a magnetic stirring bar.
2. Add 1.5 mL of *N,N*-dimethylacetamide (DMAc) to dissolve them.
3. Deoxygenate the above solution by purging with argon for 15 min and flame-seal the ampule (*see* **Note 4**).
4. Heat the mixture at 70 °C for 24 h.
5. Cool the ampule with ice water bath.
6. Open the ampule, pour the reaction mixture into an excess of methanol to precipitate, and collect the precipitate by filtration.
7. Dissolve the precipitate with distilled water and dialyze (membrane cutoff of 3500 or 1500 g mol^{-1}, depending on the estimated molecular weight of glycopolymer.) against distilled water for 2 days to remove unreacted monomer and impurities, and then freeze-dry for 3 days to obtain the polymer as a pink powder, yield of the product ~50 % (*see* **Notes 5** and **6**).

3.3 Preparation of Spindle Fe_2O_3 Nanoparticles (NPs)

1. To a solution of 54 g of $FeCl_3 \cdot 6H_2O$ in 100 mL of deionized water in a 250 mL Pyrex bottle, add NaOH solution (21.6 g NaOH pellets in 100 mL of deionized water) drop wise under stirring (*see* **Note 7**).
2. After completion of the exothermic reaction, a dark brown sticky liquid is obtained.
3. Continue stirring for 5 min until the formed material evenly distributes.

4. Remove the stirrer, place the mixture in an oven at 100 °C and leave undisturbed for 7 days.
5. Collect the spindle iron oxide nanoparticles by repeated centrifugation, wash with deionized water, and then dry in vacuum.

3.4 Synthesis of DMA-coated Fe_2O_3 NPs

1. Add and disperse Fe_2O_3 NPs (15 mg) and DMA (250 mg) in 3 mL of *N,N*-dimethylformamide (DMF), and then treat it with a sonifier for 24 h (*see* **Note 8**).
2. Rinse the particles with ethanol (3 × 5 mL) for three times and then dry in a vacuum oven at room temperature.
3. Tan solid is obtained, yield ~16.5 mg.

3.5 Synthesis of PMAG-coated Fe_2O_3 NPs by Thiol-ene Reaction

3.5.1 Reduction of the End Groups of PMAG

1. Dissolve PMAG (100 mg, 0.0063 mmol) in 10 mL of dry DMF, stir to dissolution under nitrogen, and add hexylamine (16.5 μL, 0.126 mmol) using a syringe.
2. Stir the reaction mixture at 50 °C under nitrogen for 24 h (*see* **Note 9**).
3. Dilute the reaction mixture with 20 mL of distilled water and dialyze against distilled water using a membrane with a molecular weight cutoff of 3500 g/mol for 2 days.
4. Freeze dry to afford the final product (thiol-terminated PMAG) as white powders. Yield, 85 %.

3.5.2 Preparation of Fe_2O_3 at PMAG

1. Place DMA-coated Fe_2O_3NPs (15 mg), hexylamine (8.3 μL), triethylamine (8.8 μL), and the above-obtained thiol-terminated PMAG (100 mg) into a round-bottom flask and dispersed in 10 mL of DMF, and then expose to sonication for 10 h.
2. Dialyze the above solution (membrane cutoff of 100 000 g/mol) against distilled water for 2 days, followed by freeze-drying to afford the product.
3. Yield ~18 mg; yellowish brown solid.

3.6 Cytotoxicity of the PMAG-coated Fe_2O_3 NPs

1. Seed HeLa cells in 96-well microtiter plates (1×10^4 cells per 100 μL culture media per well) with PMAG-coated Fe_2O_3NPs at concentrations ranging from 1 to 250 μg/mL. Untreated cells are used as the negative control.
2. Incubate the plate in an incubator at 37 °C in 5 % CO_2 for 24 h.
3. Add 10 μL of CCK-8 reagent to each well, and incubate the cells for 4 h (*see* **Notes 10** and **11**).
4. Record the 450/650 nm absorbance (450 nm for soluble dye and 650 nm for viable cells) using microplate reader.

3.7 Cellular Uptake of PMAG-coated Fe_2O_3 NPs

1. Incubate HeLa cells with PMAG-coated Fe_2O_3 NPs for 6 h in Dulbecco's Minimum Essential Media (DMEM) plus 10 % fetal bovine serum in Petri dish.
2. Take the samples out from the incubator and pour away the upper medium.
3. Use enzyme trypsin to detach the cells from the Petri dish surface, transfer the media to a 15 mL centrifuge tube.
4. Add PBS7.4 buffer to a total volume of about 10 mL, centrifuge (179 ×*g*, 5 min), and pour away the upper medium to remove any extracellular particles.
5. Add quantitative volume of pH 7.4 PBS buffer (*V*) to homogenize the cells, and measure the concentration of Fe (*C*) using the ICP-OES technique.
6. Calculate the number of uptaken particles per cell.

The number of nanoparticles (*N*) uptaken per cell is calculated from the equation

$$N = \frac{C \cdot V / w}{D \cdot p} / U$$

U: the number of cells obtained using a plate count

D: the volume of iron oxide nanoparticles

p: the density of iron oxide nanoparticles

w: the mass fraction of iron in the ferric oxide

4 Notes

1. Use a syringe to take methacryloyl chloride, add dropwise through a needle slowly, and avoid contact with moisture and amines.
2. High temperature should be avoided for removing solvents using rotary evaporation; the preferred temperature is 30 °C.
3. The monomer should be kept below 4 °C and avoid contact with moisture.
4. An ampule with rubber septum could be used instead; in this case, there is no need to flame seal.
5. The ^{1}H NMR of PMAG are given bellow. ^{1}H NMR (300 MHz, D_2O, ppm): 7.3–8.1 (m, 7H, CH of naphthalene units), 5.0–5.35 (d, 1H, H-1 of glucose), 3.35–3.95 (m, 6H, H-2, H-3, H-4, H-5, H-6 of glucose), 1.7–1.8 (s, 2H, CH_2 of main chain), 0.8–1.0 (s, 3H, CH_3 of main chain).

6. The M_n value of PMAG can be determined by ^{1}H NMR, based on the following equation, $M_{n(NMR)} = [(I_{(3.35-3.95)}/6)/(I_{(7.3-8.1)}/7)] \times M_M + M_{CTA}$, where $I_{(7.3-8.1)}$ and $I_{(3.35-3.95)}$ represent the integration of $7 \times$H of the naphthalene ring and $6 \times$H of the sugar moiety.
7. The stirring should be vigorous and fast.
8. The amplitude of sonifier could be lower, for example 20 %, to make sure that the NPs disperse well, while the temperature of the mixture wouldn't increase too much.
9. DMA-coated Fe_2O_3 particles can be directly added to the mixture containing PMAG, hexylamine and triethylamine for the preparation of PMAG-coated Fe_2O_3 NPs in one-pot, instead of isolating the thiol-terminated PMAG.
10. Light should be avoided and bubbles should also be avoided when adding CCK-8.
11. Shake the plate gently after adding CCK-8 reagent to ensure thorough mixing.

Acknowledgments

This work was supported by the National Natural Science Foundation of China (No. 21374069).

References

1. Boisselier E, Astruc D (2009) Gold nanoparticles in nanomedicine: preparations, imaging, diagnostics, therapies and toxicity. Chem Soc Rev 38:1759–1782
2. Barua S, Yoo JW, Kolhar P et al (2013) Particle shape enhances specificity of antibody-displaying nanoparticles. Proc Natl Acad Sci 110:3270–3275
3. Harrington MJ, Masic A, Holten-Andersen N et al (2010) Iron-clad fibers: a metal-based biological strategy for hard flexible coatings. Science 328:216–220
4. Li X, Bao M, Chen G et al (2014) Glycopolymer-coated iron oxide nanoparticles: shape-controlled synthesis and cellular uptake. J Mater Chem B 2:5569–5575
5. Ting SRS, Min EH, Stenzel M et al (2010) Controlled/living *ab initio* emulsion polymerization via a glucose RAFT *stab*: degradable cross-linked glyco-particles for Concanavalin A/*Fim*H conjugations to cluster *E. coli* bacteria. Macromolecules 43:5211–5221
6. Lu J, Zhao Y, Chen G (2014) One-pot synthesis of glycopolymer-porphyrin conjugate as photosensitizer for targeted cancer imaging and photodynamic therapy. Macromol Biosci 14: 340–346
7. Zhang Z, Zhu X, Zhu J et al (2006) Thermal-initiated reversible addition-Fragmentation Chain transfer polymerization of methyl methacrylate in the presence of Oxygen. J Polym Sci A Polym Chem 44:3343–3354
8. Lee H, Lee BP, Messersmith PB (2007) A reversible wet/dry adhesive inspired by mussels and geckos. Nature 448:338–341

Chapter 13

Synthetic Approach to Glycopolymer Base Nanoparticle Gold(I) Conjugate: A New Generation of Therapeutic Agents

Christian K. Adokoh, James Darkwa, and Ravin Narain

Abstract

Advances in nanotechnology have led to the fabrication of nano-constructs of organic or inorganic origins with well-defined structures, surface properties, and can be made to respond to physical or chemical stimuli. These nano-constructs can provide a shift in the way diagnostic and therapeutic drugs are delivered to achieve target specificity and increased retention of therapeutic doses for considerable improvement in the overall treatment of the tumors. In this case we describe here a synthetic approach to glycopolymer base nanoparticle gold(I) conjugate for cancer therapy.

Key words Glycopolymer, Gold(I) triphenylphosphine, Target specificity, Glyconanoparticles

1 Introduction

Cancer is a disease caused by aberrant cell cycle progression and defective apoptosis induction due to the activation of proto-oncogenes and/or inactivation of tumor suppressor genes [1]. Cancer is the leading cause of death worldwide, accounting for 8.2 million deaths in 2012 [2,3]. The burden is increasing in economically developing countries as a result of population aging and growth as well as, increasingly, an adoption of cancer-associated lifestyle choices which include smoking, physical inactivity, and "westernized" diets [4]. It was estimated in 2012 that 14.1 million new cancer cases and 32.6 million people living with cancer (within 5 years of diagnosis) worldwide [5]. The current treatment of cancer involves radiation, chemotherapy, hormone ablation, and radical prostatectomy, which are associated with significant toxicity and recurrence risk [5,6]. To improve the outcome of patients with cancer, there is a need for new therapies, for higher cure rates while minimizing treatment-related toxicities. Advances in nanotechnology have led to the fabrication of nano-constructs of organic or inorganic origins with well-defined structures, surface properties, and can be made to respond to physical or chemical

Xue-Long Sun (ed.), *Macro-Glycoligands: Methods and Protocols*, Methods in Molecular Biology, vol. 1367,
DOI 10.1007/978-1-4939-3130-9_13, © Springer Science+Business Media New York 2016

stimuli. These nano-constructs can provide a shift in the way diagnostic and therapeutic drugs are delivered to achieve target specificity and increased retention of therapeutic doses for considerable improvement in the overall treatment of the tumors. The evolving molecular events often provide the intervening candidate targets for the development of cancer therapy. Some of the most promising targets are ASGP receptors and p53, a well-established and frequently mutated tumor suppressor in human cancer. p53 for example plays a critical role in tumor suppression mainly by inducing growth arrest, apoptosis, and senescence, as well as by blocking angiogenesis. In addition, p53 generally confers the cancer cell sensitivity to chemo radiation. Thus, p53 and ASGP becomes the most appealing target for mechanism-driven anticancer drug discovery [7]. Therefore, our motivation is towards the development of novel glycopolymer nanomedicine for targeting ASGP receptors overexpressed on the liver cancer cellular surface. Glycopolymers and glyconanoparticles are of interest here as they expected to be safe nano-constructs for the safe delivery of the cancer drug ($AuPPh_3$). We have suggested three steps strategies to incorporate thiol and dithiocarbamate functionality for the stabilization of gold nanoparticles and the cancer drug for therapeutic application via RAFT polymerization method [8].

The first step is the fabrication of statistical glyco-glycidyldithiocarbamate polymers: p(GMA-EDAdtc-st-LAEMA) and p(GMA-EDAdtc-st-GAEMA) (GPdtc) (Fig. 1). This involves a three-step approach: (1) the statistical polymerization of gluconamidoethyl methacrylate and an epoxide (glycidyl methacrylate) (method section, subheading 1), (2) post-decoration of the epoxide with oligoamine to yield cationic copolymers P(GMA-EDA)-st-P(GAEMA) and P(GMA-EDA)-st-P(LAEMA) [9], and (3) the resultant cationic glycopolymer bearing terminal amino group contains primary amino groups as our functionalization targeted moiety to react with the activated carbon disulfide at 0 °C (Subheading 3) to give statistical glyco-glycidyldithiocarbamate polymers (Fig. 1).

In the second approach glyco-dithiocarbamate stabilized gold nanoparticles were synthesized via photoirradiation method using Irgacure 2959 as photoinitiator (Fig. 2). Finally, a spherical glyco-dithiocarbamate stabilized gold nanoparticles were conjugated to gold triphenylphosphine as described in Fig. 2.

2 Materials

2.1 Chemicals

1. 4-Cyanopentanoic acid dithiobenzoate (CTP).
2. Carbon disulfide.
3. Ethylene diamine (EDA).

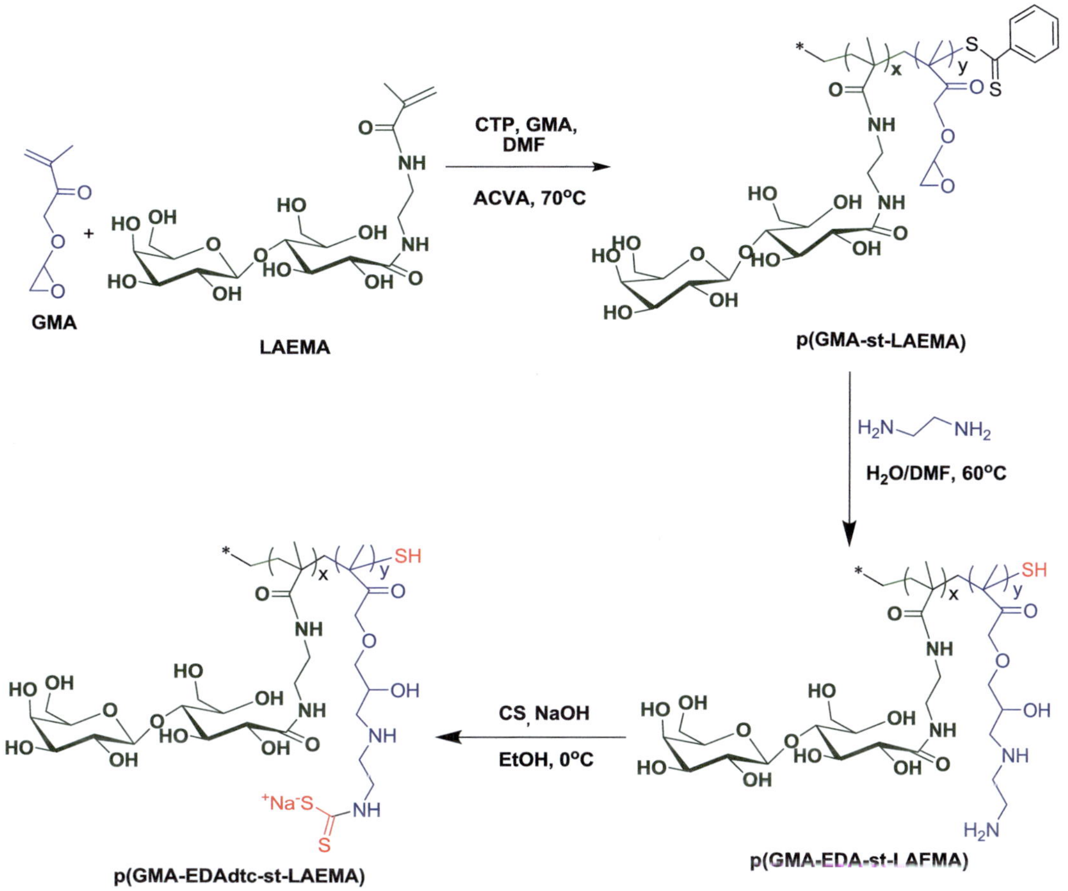

Fig. 1 Synthesis of statistical dithiocarbamate copolymer p(GMA-EDAdtc-st-LAEMA) (LPdtc)

4. Tetrahydrothiophene.
5. Hydrogen tetrachloroaurate.
6. Triphenyl phosphine.
7. Glycidyl methacrylate (GMA, 97 %, Aldrich) is purified by column chromatography using alumina prior to its use.
8. 4,4-Azobis (4-cyanovaleric acid) (ACVA) (Acros Organics).
9. 2-Gluconamidoethyl methacrylamide hydrochloride (GAEMA) and 2-lactobionamidoethyl methacrylamide (LAEMA) were synthesized as described in [10–13].

2.2 Buffer and Other Solutions

1. Doubly distilled deionized water is used in all experiments.
2. PBS buffer solution (pH 7.4) 1.0 M.
3. Buffer solution for GCP: sodium acetate 0.5 M/acetic acid.
4. Six near-monodisperse PEO standards (Mp) 1,010–101, 200 g/mol).

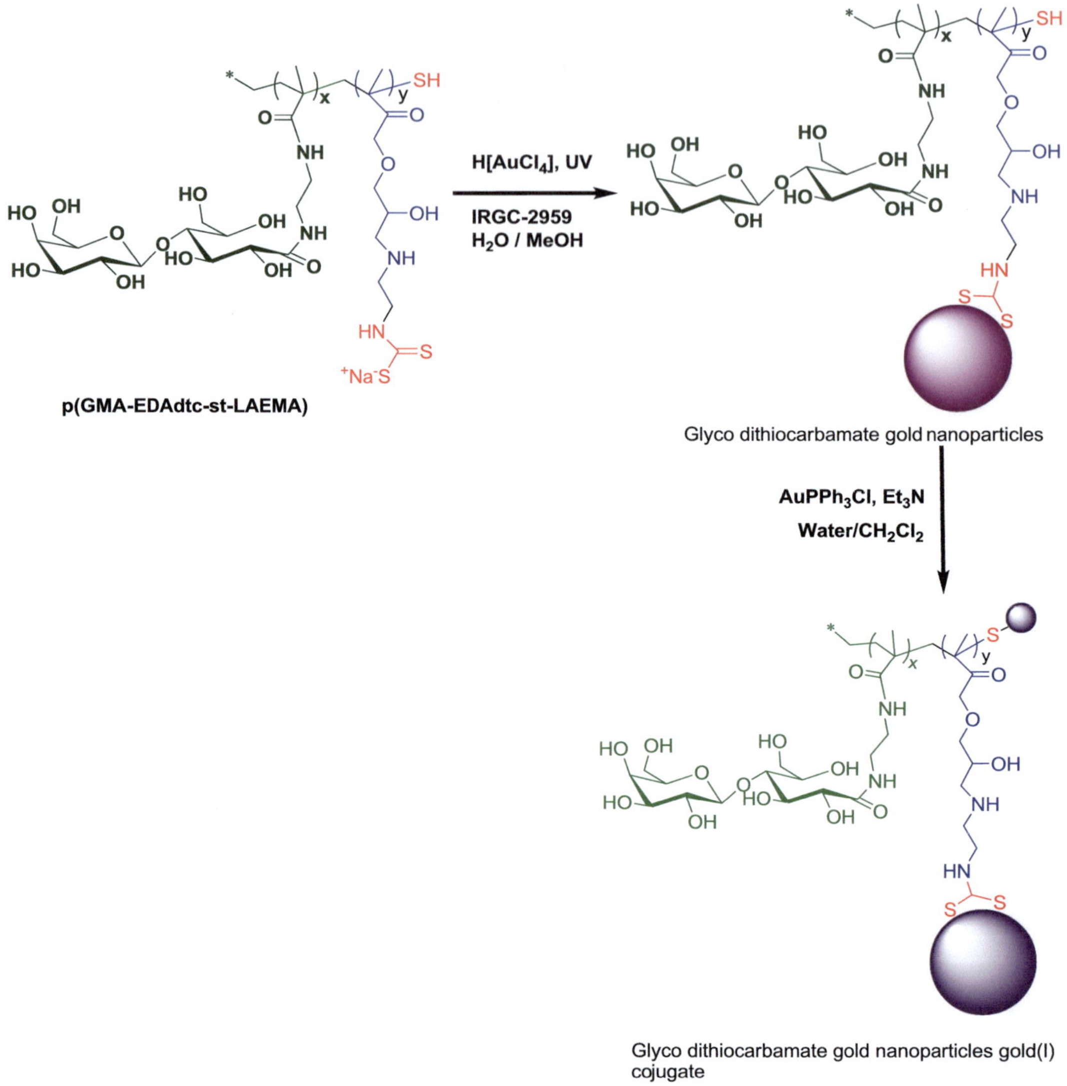

Fig. 2 Synthesis of polymer-functionalized gold nanoparticles and their conjugation to the anticancer drug, Au(1)PPh$_3$

5. 10 % NaOH (1.7 M) solution for critical flocculation concentration (CFC) measurement.

2.3 NMR Spectroscopy

^{1}H NMR spectra of the monomers and polymers are recorded using a Varian spectrometer (500 MHz).

2.4 Chromatography

2.4.1 Preparative GPC

Aqueous gel permeation chromatography analysis is performed on a conventional Viscotek Instrument using 0.5 M sodium acetate/0.5 M acetic acid buffer as eluent, two Waters WAT011545 columns at room temperature and a flow rate of 1.0 mL/min using 0.5 M sodium acetate/0.5 M acetic acid buffer as eluent.

2.4.2 Analytical GPC

1. Seven near-monodisperse Pullulan standards (M_w 5900–404,000 g/mol) are used for calibration.
2. The products are analyzed by an Agilent HPLC 1100 interfaced with an Electro-Spray Ionization Agilent Mass Spectrometer Model 6120 with a Chemstation data system LCMSD B.03.01.

2.5 Fourier Transform Infrared

Fourier transform infrared (FT-IR) spectral analyses using KBr pellets of the synthesized samples are carried out on a Nicolet 8700 (Thermo) instrument and the diffuse reflectance spectra are scanned over a range of 4000–400 cm^{-1} wave numbers.

2.6 Dynamic Light Scattering

1. Dynamic light scattering (DLS) measurements are performed with a ZetaPlus-Zeta Potential Analyzer (Brookhaven Instruments Corporation) at a scattering angle $\theta = 90°$.
2. p(GMA-EDA-st-LAEMA)-stabilized gold nanoparticles solutions are filtered through Millipore membranes (0.45 μm pore size).
3. The data are recorded with Omni size software.

2.7 UV–Visible Spectroscopy

UV–visible absorption spectra (400–800 nm) were recorded on a Cary UV 100 spectrophotometer from the aqueous solutions of polymeric AuNPs at room temperature.

2.8 Photoirradiation

The photo irradiation of the reaction mixture was carried out using 16, 12, 8, and 4 75 W UV lamps at a wavelength of 300 nm in a Rayonet photo reactor (Southern N.E. Ultraviolet Co.).

3 Methods

The RAFT polymerization approach and two-step strategies were used to prepare stable surface-functionalized gold(I) dithiocarbamate gold nanoparticles conjugate. Firstly, RAFT statistical copolymers of glycidyl methacrylamide (GMA) and 2-gluconamidoethyl methacrylamide, P(GMA-st-GAEMA), glycidyl methacrylamide, and 2-lactobionamidoethyl methacrylamide P(GMA-st-LAEMA) were synthesized employing ACVA as the initiator and CTP as the chain transfer agent, according to previously reported procedures (*see* **Note 1**) [8,14,15] as described in details below. This was followed by post-decoration of the glycidyl methacrylamide epoxide surface with ethylene diamine (EDA) to yield cationic copolymers: $P(GMA\text{-}EDA)_7$-st-$P(GAEMA)_{22}$ and $P(GMA\text{-}EDA)_{8\text{-st-}P}(LAEMA)_{24}$ with molar masses of *ca.* 20 and 22 kDa, respectively, with narrow molecular weight distributions ($M_w/M_n < 1.33$). In this step the terminal dithioester RAFT group was reduced to free thiol by aminolysis during the amine decoration (Subheading 3) [16] (*see* **Note 2**). This thiol group was targeted as the stabilizing agent of the triphenylphosphine as the dithiocarbamate moiety was believed to stabilize the gold nanoparticles or vice versa. Lastly, the DTC polymers were synthesized by mixing EDA-decorated polymers and carbon disulfide (*see* **Note 3**) and were characterized by ^{1}H NMR and FTIR spectroscopy. The average hydrodynamic diameter of dithiocarbamate polymers in 150 mM PBS was determined by DLS. Average sizes of 27.0 and 44.7 nm with moderate PDI (~0.30) were recorded [17,18]. The net charge of polymeric dithiocarbamate derived from LAEMA and GAEMA were slightly negatively charged (−10.13 and −5.00), suggesting that all free amines in the polymers were functionalized with a dithiocarbamate motif.

The second step is the glyco-dithiocarbamate polymer-coated gold nanoparticles synthesis via photo irradiation using Irgacure-2959 (IRGC) as a photoinitiator (*see* **Note 4**) and was confirmed by transmission electron microscopy (TEM), UV–Vis spectroscopy, and DLS analysis.

The size distributions of glyco-dithiocarbamate polymer-coated gold nanoparticles were relatively narrow with polydispersity index (PDI) between 0.17 and 0.28 and the particle size distributions as measured by TEM and DLS for all the GNPs ranged between 12.5 ± 3.7 and 45.8 ± 12.7 nm (Fig. 5) [18]. The stability of glyco-dithiocarbamate polymer-modified gold nanoparticles was assessed by the dispersion of nanoparticles in high salt and pH conditions. No shift of the UV–Vis peak was observed before and after GNPs suspension in 10 % NaCl solution, even after 1 week at normal temperature an indication that GNPs are stable in physiological conditions (Fig. 3a). The gold nanoparticles were also found to be reasonably stable at physiological pH, as a bathochromic shift of ~3 nm was observed at pH values between 6 and 8, with no broadening of surface plasmon resonance (SPR) band (Fig. 3b) [18].

Finally, a biologically active glyco-decorated gold nanoparticles gold(I) conjugates were prepared using glucose-derived and galactose-decorated dithiocarbamate-stabilized gold nanoparticles of about 13 nm in size and gold(I) triphenylphosphine chloride in a biphasic medium (Fig. 2). The successful conjugation of gold(I) was followed by TEM, UV–visible spectroscopy, DLS, and physical appearance. The UV–Vis spectra showed a red shift of absorption band after the conjugation of gold(I) onto the surface of polymeric galactose gold nanoparticles suggesting an aggregation of particles after conjugation which was evidenced in the TEM micrograph (Fig. 5). This was also evidenced by the change of color from red to purple (Fig. 4) after conjugation. The glycopolymer functionalized gold(I) conjugate revealed an increase in size (e.g., 13.0–21.4 nm), which confirms the successful conjugation of the glycopolymer to the gold triphenyl phosphine (Fig. 5c, d) [18].

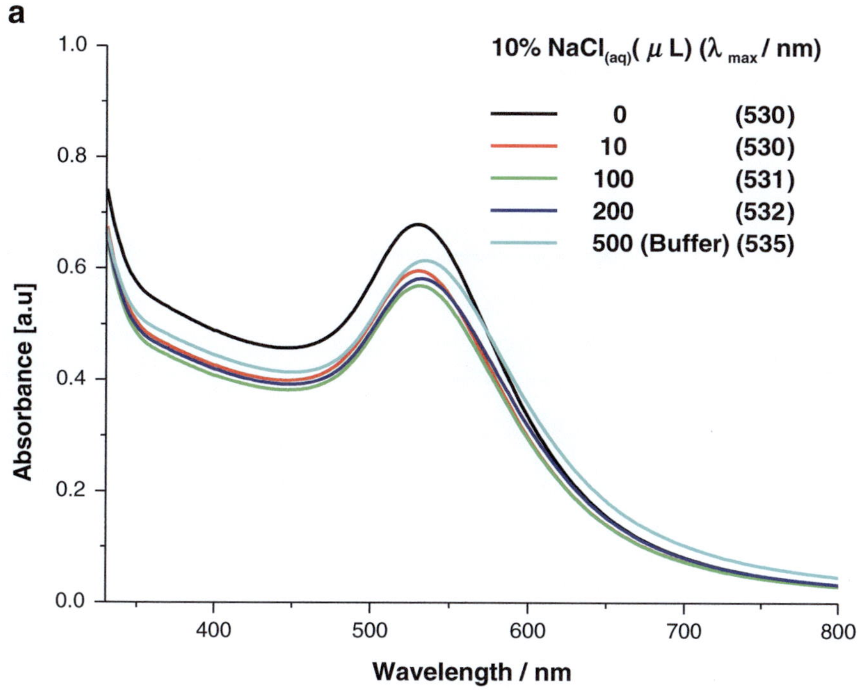

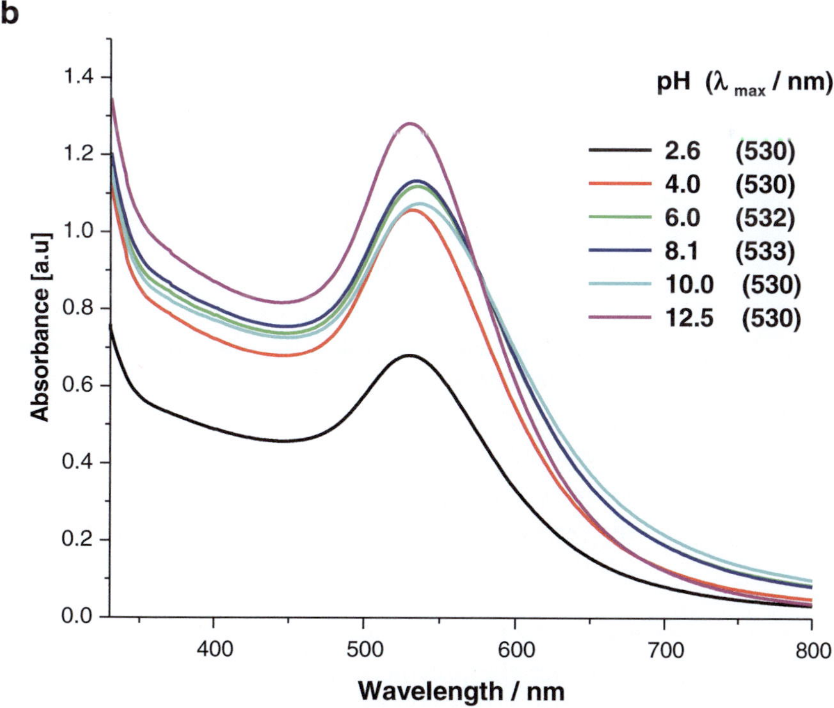

Fig. 3 Representative UV–Vis spectra of (**a**) gold nano particles after their dispersion in high-salt solution of 10 % NaCl and (**b**) gold nanoparticles at various pH using glyco-dithiocarbamate polymer-coated gold nanoparticles [P(GMA-EDAdtc-st-LAEMA)]AuNP [18]

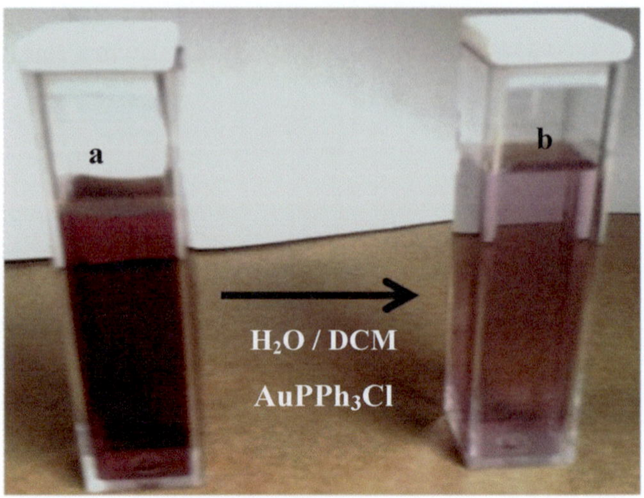

Fig. 4 Synthesis of galactose dithiocarbamate-coated gold nanoparticles gold(I) conjugate (**a**) EDAdtc-st-LAEMA)]AuNP, (**b**) [P(GMA-EDAdtc(Au(I)PPh_3)-st-LAEMA)]AuN [18]

3.1 Synthesis of Glyco-Decorated Dithiocarbamate Statistical Polymers

1. In a 10 mL flask, dissolve LAEMA (1.00 g, 2.04×10^{-3} mol) and GMA (0.17 g, 1.17×10^{-3} mol) in doubly distilled water (4.5 mL).
2. Add a solution of ACVA (3.5 mg, 0.0125 mmol) and 4-cyanopentanoic acid dithiobenzoate (CTP) (17.8 mg, 0.062 mmol) in N,N'-dimethylformamide (DMF) (1 mL).
3. Degas the mixture via three freeze-pump-thaw cycles for 30 min and seal with parafilm.
4. Place the reaction flask to a pre-heated oil bath at 80 °C and stir the mixture for 24 h.
5. Placed the reaction flask into liquid nitrogen to quench the polymerization.
6. Pour reaction mixture into diethyl ether precipitate out the polymer.
7. Dissolve the precipitated polymer into 5.0 mL of DMF and dialyze against 4 L double distilled water, replace three times per day over the course of 3 days (using pore size 12 kDa cutting molecular weight to remove unreacted monomer) and freeze dry to yield white solid.
8. Conduct the polymerization with a second monomer, GMA (0.166 g, 1.17×10^{-3} mol) and GAEMA (1.07 g, 3.5×10^{-3} mol) in water (5 ml), ACVA (3.5 mg, 0.0125 mmol), and 4-cyanopentanoic acid dithiobenzoate (CTP) (17.4 mg, 1.17 mmol) in 1 mL 1,4-dioxane also treated as steps above to yield second polymer as pink solid.

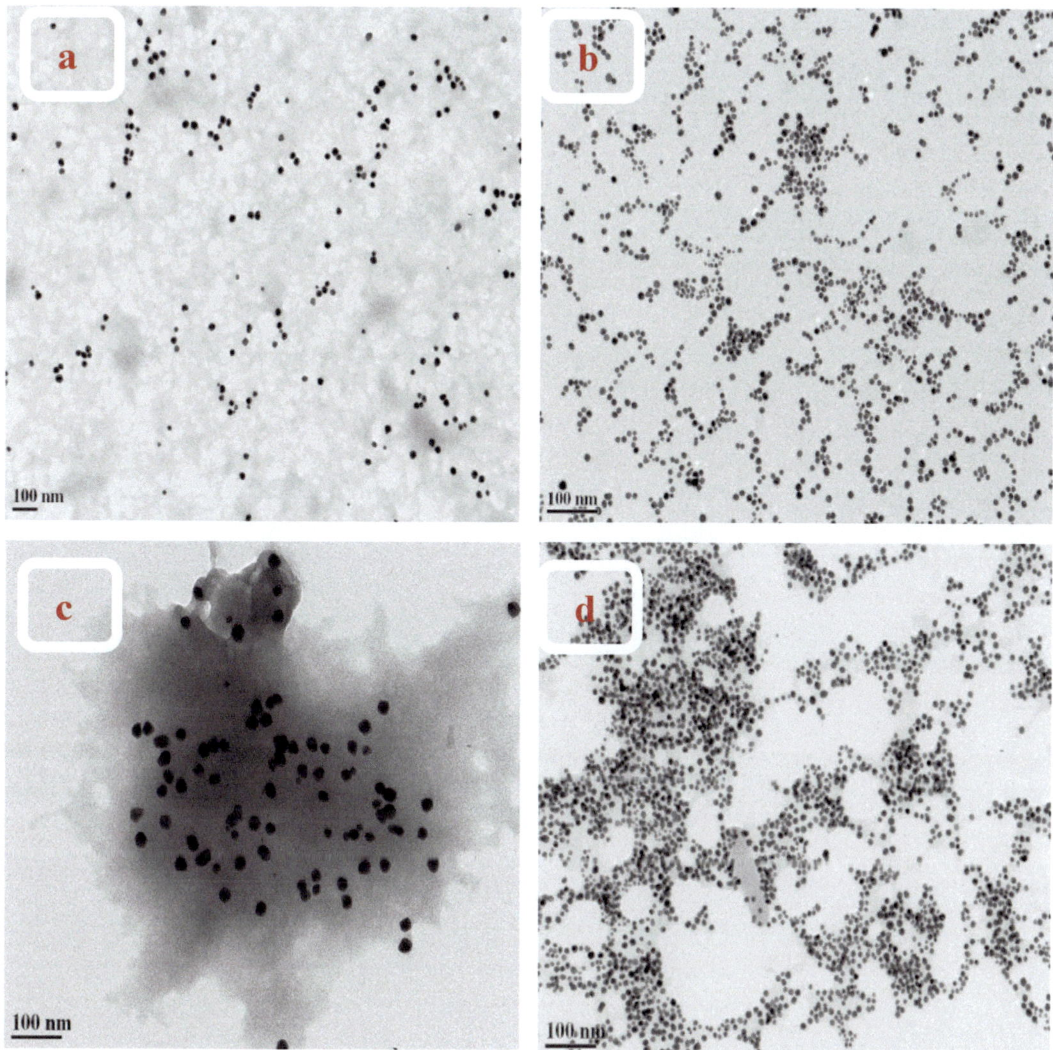

Fig. 5 Representative TEM micrograph of glyco-dithiocarbamate polymer-coated gold nanoparticles: (**a**) EDAdtc-st-LAEMA)]AuNP, (**b**) [P(GMA-EDAdtc-st-GAEMA)]AuNP and glyco-decorated gold nanoparticles gold(I) conjugates: (**c**) [P(GMA-EDAdtc(Au(I)PPh_3)-st-LAEMA)]AuNP, (**d**) [P(GMA-EDAdtc(Au(I)PPh_3)-st-GAEMA)]AuNP [18]

3.2 Decoration of Statistical Copolymers by Ethylenediamine (EDA) [p(GMA-EDA)-st-p(LAEMA)]

1. Dissolve EDA (2.0 mL, 10.5 mmol) in anhydrous DMF (10 mL) in a 25 mL round-bottom flask equipped with a magnetic stirring bar.
2. Add the respective polymers (50 mg) in 2 mL of DMF drop wise.
3. After reaction at 60 °C for 7 h, dialyze the reaction mixture (pore size 12 kDa cutting molecular weight) against 4 L of distilled water, exchange water three times per day over the course of 3 days.
4. Finally, freeze-dry to afford the amine decorated polymers for further characterization.

3.3 Synthesis of Polymeric DTC Polymers

1. Dissolve EDA decorated polymers (100 mg) in 0.25 M NaOH ethanolic solution (8.0 mL) and cool the solution to 0 °C.
2. Add carbon disulfide (0.04 mL, 0.71 mmol) in ethanol (2 mL) in drops wise.
3. Stir the reaction mixture for about 4 h at 0 °C and allow the mixture to stir overnight to afford the product.
4. Collect the solid by filtration and wash with ethanol to remove excess starting materials.
5. Dry the solid under vacuum to afford a degree of colored products, depending on monomer and molecular weight targeted. p(GMA-EDAdtc)-st-p(LAEMA) (LPdtc).

3.4 Synthesis of Statistical Glucose-Derived and Galactose Copolymer-Coated Au Nanoparticles in the Presence of Irgacure-2959 (IRGC)

1. Dissolve polymers: LPdtc (ca. 22 kDa) and GPdtc (ca. 20 kDa) in deionized water (15 mL) via sonication.
2. Add the solution to $HAuCl_4$ (0.005 mg/mL) solution in double-distilled water (5 mL).
3. Add a methanolic solution of Irgacure-2959 (8.5 mg, 0.038 mmol) initiator (5 mL) to $HAuCl_4$ and polymer mixture and degas via three freeze-pump-thaw cycles for 15 min
4. Conduction irradiation with photo reactor using sixteen, twelve, eight, and four 75 W UV lamps at wavelength 300 nm in a Rayonet photo reactor (Southern N.E. Ultraviolet Co.) for 15 min.
5. Take off the reaction mixture from the photo reactor and stir at room temperature for 5 min.
6. Collect the gold nanoparticles produced by filtering through Millipore membranes (0.45 μm pore size) and by centrifugation at 47,850 × *g* force for 2 h to remove excess polymer.
7. Re-suspend the pellet in deionized water and stored in a fridge for further characterization and use.
8. Three forms of polymer concentrations (1, 2, and 5 mg) gold nanoparticles are synthesized to study concentration variation effect.

3.5 Synthesis of Statistical Copolymer AuNP-Decorated Gold(I) Triphenylphosphine Complex

1. Add a solution of $[AuCl(PPh_3)]$ (2 mg, 0.004 mmol) and triethylamine (8.5 μL) in dichloromethane (5 mL) to a solution of respective polymeric modified galactose dithiocarbamate-stabilized AuNPs in water (6 mL; ca. 53.5 nM) under nitrogen.
2. Stir the resulting solution for 3 h at room temperature (23 °C).
3. Separate the aqueous layer and centrifuge at 47,850 × *g*-force for 2 h three times to remove unwanted materials
4. Resuspend the pellet in water for further characterization.

3.6 Critical Flocculation Concentration

1. Centrifuge the filtered glyco-stabilized gold nanoparticles at 26,916 × *g*-force for 90 min at room temperature (23 °C).
2. Resuspend the modified particles in 1.0 M phosphate buffer solution.
3. Centrifuge three times to ensure complete removal of free ligands.
4. After each 5 min, add 500 μL of 1.0 M phosphate buffer as well as 10, 100, and 200 μL of 10 % NaCl (1.7 M) solutions to each solution and allowed standing for 1 h.
5. The approximate glyco-stabilized gold nanoparticles concentration of *ca.* 53.4 nM (3 mL) is used for critical flocculation concentration (CFC) test.

4 Notes

1. The RAFT polymerization is achieved at 80 °C.
2. In the preparation of amine-decorated glycopolymer the ethyl diamine (EDA) is in a 30-fold molar excess and the reaction is carried out at 60 °C.
3. The functionalization of amine glycopolymers is done under 0 °C and all values are based on the amount of the P(GMA) in the polymer.
4. In the synthesis of glyco gold nanoparticles the molar ratios of $HAuCl_4$/IRGC in double-distilled water is 1:3 as polymer concentration is varied.

References

1. Hanahan D, Weinberg RA (2000) The hallmarks of cancer. Cell 100:57–70
2. Ferlay J, Soerjomataram I, Ervik M, Dikshit R, Eser S, Mathers C, Rebelo M, Parkin DM, Forman D, Bray F (2013) Cancer incidence and mortality worldwide: IARC cancer base No. 11 [Internet]. Lyon, France: International Agency for Research on Cancer; GLOBOCAN 2012 v 1.0
3. Bray F, Ren JS, Masuyer E, Ferlay J (2013) Global estimates of cancer prevalence for 27 sites in the adult population. Int J Cancer 132:1133–1145
4. Jemal A, Bray F, Center MM, Ferlay J, Ward E, Forman D (2011) Global cancer statistics. Cancer J Clin 61:69–90
5. Viani GA, Pellizzon AC, Guimaraes FS, Jacinto AA, Dos Santos Novaes PE, Salvajoli JV (2009) High dose rate and external beam radiotherapy in locally advanced prostate cancer. Am J Clin Oncol 32:187–190
6. Yoshioka Y (2009) Current status and perspectives of brachytherapy for prostate cancer. Int J Clin Oncol 1:31–16
7. Wang Z, Sun Y (2010) Targeting p53 for novel anticancer therapy. Transl Oncol 3:1–12
8. Ahmed M, Narain R (2013) Applications of bioconjugates. Chem Bioconjug Appl Bioconjug: 444–456
9. Chang CW, Bays E, Tao L, Alconel SNS, Maynard HD (2009) Differences in cytotoxicity of poly(PEGA)s synthesized by reversible addition-fragmentation chain transfer polymerization. Chem Commun 24:3580–3582
10. Deng Z, Li S, Jiang X, Narain R (2009) Well-defined galactose-containing multi-functional

copolymers and glyconanoparticles for biomolecular recognition processes. Macromolecules 42:6393–6405

11. Narain R, Armes SP (2002) Synthesis of low polydispersity, controlled chemistry. Chem Commun 23:2776–2777
12. Narain R, Armes SP (2003) Direct synthesis and aqueous solution properties of well-defined cyclic sugar methacrylate polymers. Macromolecules 36:4675–4678
13. Narain R, Armes SP (2003) Synthesis and aqueous solution properties of novel sugar methacrylate-based homopolymers and block copolymers. Biomacromolecules 4: 1746–1758
14. Ahmed M, Narain R (2012) The effect of molecular weight, compositions and lectin type on the properties of hyperbranched glycopolymers as non-viral gene delivery systems. Biomaterials 33:3990–4001
15. Ahmed M, Deng Z, Liu S, Lafrenie R, Kumar A, Narain R (2009) Cationic glyconanoparticles: their complexation with DNA, cellular uptake, and transfection efficiencies. Bioconjug Chem 20:2169–2176
16. Wilson-Welder JH, Torres MP, Kipper MJ, Mallapragada SK, Wannemuehler MJ, Narasimhan B (2009) Vaccine adjuvants: current challenges and future approaches. J Pharm Sci 98:1278–1316
17. Ahmed M, Mamba S, Yang X-H, Darkwa J, Kumar P, Narain R (2013) Synthesis and evaluation of polymeric gold glyco-conjugates as anticancer agents. Bioconjug Chem 24:979–986
18. Adokoh CK, Quan S, Hitt M, Darkwa J, Kumar P, Narain R (2014) Synthesis and evaluation of glycopolymeric decorated gold nanoparticles functionalized with gold-triphenyl phosphine as anti-cancer agents. Biomacromolecules 15:3802–3810

Chapter 14

Multivalent Glycopolymer-Coated Gold Nanoparticles

Sarah-Jane Richards, Caroline I. Biggs, and Matthew I. Gibson

Abstract

Glycosylated noble metal nanoparticles are a useful tool for probing biological binding events due to their aggregation-induced color changes, particularly for lectins that have multiple binding sites. To overcome the challenges of colloidal instability, which leads to false-positive results, it is essential to add polymeric coatings to these particles. Here we describe a versatile, and reliable, approach to enable coating of gold nanoparticles using well-defined polymers, with carbohydrate end groups. This produces multivalent nanoparticles that are both colloidally stable, but still retain their rapid colorimetric responses to lectin binding.

Key words Gold nanoparticles, Glycopolymers, RAFT polymerization, Lectins

1 Introduction

Carbohydrate-functionalised AuNPs (glycoAuNPs) are emerging as important tools for the colorimetric determination of carbohydrate-protein interactions as they constitute a good biomimetic model of carbohydrates at the cell surface (glycocalyx). GlycoAuNPs are attractive biosensors due to the multivalent presentation of carbohydrates can compensate for the low affinity of individual protein-carbohydrate interactions and the inherent multivalency of lectins should lead to a colorimetric response, as a result of aggregation, if the correct lectin-carbohydrate paring is present, changing from red to blue (Fig. 1). This therefore provides a promising platform for new sensors, without the need for expensive equipment and fluorescent- or radio-labeling of proteins.

The colorimetric change associated with the aggregation of AuNPs using protein-carbohydrate interactions has been exploited in the detection of lectins [1–4], bacteria [5, 6] and different strains of influenza [7]. We have shown that mannosylated glycoAuNPs can be used to distinguish between bacteria with and without type 1 fimbriae, but that the mode of presentation of the carbohydrates and their distance from the gold surface impacts greatly on the

Xue-Long Sun (ed.), *Macro-Glycoligands: Methods and Protocols*, Methods in Molecular Biology, vol. 1367, DOI 10.1007/978-1-4939-3130-9_14, © Springer Science+Business Media New York 2016

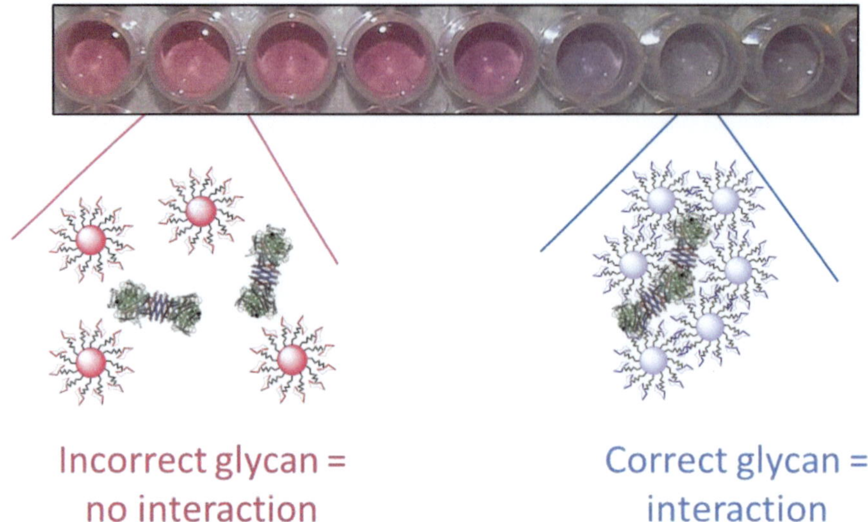

Fig. 1 Overview of the approach used to allow direct measurement of carbohydrate-lectin interactions by monitoring a colorimetric change (*red* to *blue*) due to aggregation of glycoAuNPs bound to lectins

detection readouts. Whilst AuNPs are appealing scaffolds, directly glycosylated nanoparticles have intrinsically low colloidal stability, which can result in either false-positive aggregation responses in biological media or can prevent high-throughput application due to large errors. Conversely, very slow responses (aggregation) are observed for particles with polymer coatings, which therefore limits their application as point-of-care diagnostics [6]. Therefore via a screening study we demonstrated that through precision macromolecular engineering using RAFT (reversible addition-fragmentation chain transfer) polymerization it is possible to both improve colloidal stability whilst avoiding the problem of presenting a steric block to aggregation [4]. The key features of this are a polymer coating of just the right length, and by using the RAFT process, a latent thiol is installed on each chain end, suitable for conjugation to the gold surface without any further chemical modification.

Herein, we describe a polymer-stabilized glycosylated gold nanoparticle platform for the high-throughput, label-free screening of carbohydrate-lectin interactions and demonstrate the potential applications within a range of glycobiological assays (Fig. 1). A mix-and-match synthetic strategy enables huge chemical space to be explored, with the outputs being read in a simple multi-well plate format by either a microplate reader, or simply using a digital camera.

Briefly, 60 nm AuNPs are functionalised with precisely engineered heterotelechelic poly *N*-hydroxyethyl acrylamide polymers bearing a carbohydrate moiety at one end for lectin interaction and a thiol at the other for gold particle attachment. These multivalent, carbohydrate-functionalized, stable AuNPs can then be used as probes for determining lectin interactions and lectin discrimination.

2 Materials

2.1 Chemicals

1. Potassium phosphate tribasic (Sigma Aldrich).
2. 1-Dodecanethiol (Sigma Aldrich).
3. Carbon disulfide (Sigma Aldrich).
4. 2-Bromo-2-methyl propionic acid (Acros).
5. Pentafluorophenol (PFP) (Alfa Aeser).
6. *N*-hydroxyethyl acrylamide (HEA) (Sigma Aldrich).
7. 4,4′Azobis(4-cyanovaleric acid) (ACVA) (Sigma Aldrich).
8. 60 nm gold nanoparticles (BBI solutions).
9. Low bind, low volume clear 96-well plates (Greiner Bio-one).

2.2 Solvents

1. Acetone (Sigma Aldrich).
2. Dichloromethane (Sigma Aldrich).
3. Dimethylformamide (Sigma Aldrich).
4. Toluene (Fischer Scientific).
5. Methanol (Sigma Aldrich).

2.3 Buffer

1. HEPES buffer: 10 mM HEPES, 0.15 M 0.05 M NaCl, 0.1 mM $CaCl_2$, and 0.01 mM $MnCl_2$ (pH 7.5) (*see* **Note 1**).
2. Acetate buffer: 100 mM acetate buffer (0.09 g acetic acid, 1.15 g sodium acetate in 100 mL milli-Q water) containing 1 mM aniline (pH 5.5).

2.4 Lectins (See Note 2)

1. Concanavalin A (Con A) (Vector Laboratories).
2. Peanut Agglutinin (PNA) (Vector Laboratories).
3. Soybean Agglutinin (SBA) (Vector Laboratories).
4. *Ricinus Communis* Agglutinin (RCA_{120}) (Vector Laboratories).
5. *Ulex Europeus* Agglutinin (UEA) (Vector Laboratories).
6. Wheat Germ Agglutinin (WGA) (Vector Laboratories).

2.5 Equipment

1. Bruker DPX-300 NMR spectrometer.
2. Varian 390-LC MDS SEC system equipped with a PL-AS RT/MT autosampler, a PL-gel 3 μm (50×7.5 mm) guard column, two PL-gel 5 μm (300×7.5 mm) mixed-D columns equipped with a differential refractive index, using DMF (with 1 mg/mL LiBr) as the eluent with a flow rate of 1.0 mL/min at 50 °C.
3. Bruker VECTOR-22 FTIR spectrometer using a Golden Gate diamond-attenuated total reflection cell.
4. BioTek SynergyTM multidetection microplate reader for absorbance spectrometry with a monochromator for full spectral

scanning. Minimum 450 and 700 nm wavelength measuring capability required.

5. DLS measurements were carried out using a Malvern Instruments Zetasizer Nano-ZS.

2.6 Software

1. Gen 5 1.11 multiple data collection and analysis software.
2. Image J (version 1.46a).

3 Methods

GlycoAuNPs were prepared using an optimised polymer coating method that we developed recently [4], which produces particles that are stable at physiological salt concentration and gives fast lectin detection/identification. We have suggested two methods for the carbohydrate functionalisation of RAFT-derived polymers (Fig. 2). RAFT polymerization was used due to this technique producing telechelic polymers with, in this case, a conjugatable pentafluorophenyl ester at one end and masked thiol (trithiocarbonate) at the other, which is necessary for the gold immobilization. Poly(*N*-hydroxyethyl acrylamide) (pHEA) is the optimal polymer for this as it has high water solubility, but also it does not display a lower critical solution temperature (LCST). Polymers with an LCST aggregate in water when heated, and this critical temperature decreases when on a particle surface [8, 9]. Other commonly used, neutral, hydrophilic polymers such as oligoethyleneglycol methacrylate (OEGMA) and hydroxyethyl methacrylate (HEMA) are not suitable for this reason. Also, pOEGMAs do not graft in high density to the particles due to their comb-like shape. Commercial telechelic linear PEGs can be used, but the choice of degree of polymerization (DP) is limited, which is important for ensuring fast colorimetric readouts and the choice of end group is limited, unlike here.

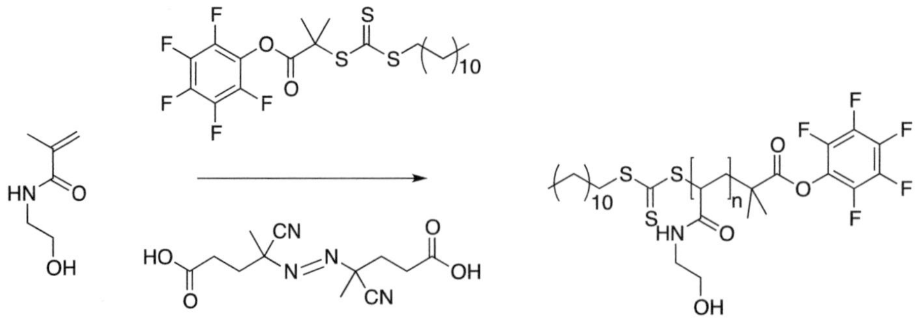

Fig. 2 Polymerization of *N*-hydroxyethyl acrylamide using 2-(dodecylthiocarbonothioylthio)-2-methylpropionic acid as the RAFT agent for simple end-group modification

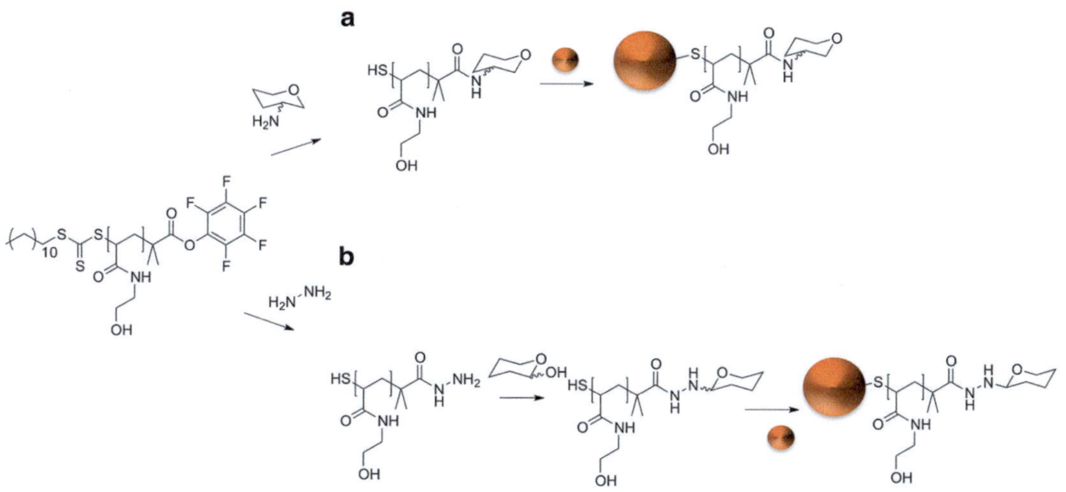

Fig. 3 Preparation of carbohydrate-functionalized gold nanoparticles. Synthetic approach resulting in polymer-stabilized gold nanoparticles with terminal carbohydrate functionality. Two synthetic approaches are suggested, (**a**) pentafluorophenol terminated polymers and amino-carbohydrates and (**b**) using hydrazide-terminated polymers with reducing sugars

To enable control over the end-groups a pentafluorophenol (PFP) trithiocarbonate RAFT agent is employed, which is an excellent mediator for acrylamides such as HEA. The PFP group is stable in the polymerization conditions but is readily converted into a range of other functionality, post-polymerization via addition of amines. This approach ensures identical polymers, with different end groups can be obtained, essential for any structure-property studies.

Two facile and versatile methods are available for the functionalization of AuNPs using native, underivatized carbohydrates with proper ligand presentation using the method presented here. This also alleviates the inconvenience and limitations of carbohydrate pre-functionalization, required for many carbohydrate conjugations, and offers rapid access to glycoAuNPs with a wide variety of glycans suitable for high-throughput microarray applications [10] as outlined in Fig. 3.

Firstly, *N*-hydroxyethyl acrylamide (HEA) is polymerized to DP 20, as detailed studies have shown this is the optimum length for 60 nm particles for saline stability and a fast readout [4]. This polymerization was carried out by RAFT using a pentafluorophenol ester α-terminated RAFT agent. This can then be functionalised directly using amino-2-deoxy-sugars (Fig. 2a) or hydrazide (Fig. 2b). The hydrazide-terminated polymers can then be functionalised with reducing sugars. The stable *N*-glycoside end-terminated polymers are formed due to the intrinsic reactivity of reducing sugars toward hydrazides. The reaction occurs under mild conditions, with good conjugation efficiencies and stereoselectivity. The reducing sugars react with α-heteroatom nucleophiles (e.g., hydrazide or alkylaminooxy groups) to give cyclic pyranosides [11].

The ω-terminal thiol from the RAFT agent can then be used to conjugate the carbohydrate-functionalised polymers, from either route, onto preformed 60 nm AuNPs. We have found that larger particles give greater responses [6]. However, the polymer coating has to be tailored to the AuNPs to be used. The post-polymerization route means that all the polymers have the same initial chain length distribution and therefore reduces the variability between particle types, but allows versatile end-group functionalization to make libraries of a variety of particles. These glycoAuNPs can then be used to test carbohydrate-protein interactions in a multi-well plate format.

3.1 Preparation of 2-(Dodecylthio-carbonothioylthio)-2-methylpropionic Acid (DMP)

1. Add 4.0 g (19.76 mmol) of dodecanethiol dropwise over 25 min to a stirred suspension of 4.2 g of K_3PO_4 (19.76 mmol) in 60 mL of acetone.
2. Add 4.1 g CS_2 (53.85 mmol). Reaction mixture turns bright yellow.
3. After 10 min, add 3 g of 2-bromo-2-methyl propionic acid (17.76 mmol). A KBr precipitate is noted.
4. Stir the reaction mixture at room temperature for 16 h.
5. Remove the solvent under reduced pressure.
6. Extract in 2 × 200 mL of dichloromethane (DCM) from 1 M HCl and wash with 200 mL of H_2O and 200 mL of Brine and dry over $MgSO_4$.
7. Concentrate under reduced pressure.
8. Purify by column chromatography using 24:75:1 ethyl acetate:petroleum ether:acetic acid ($R_{f\,product} = 0.25$).
9. Characterize the product using IR, ^1H, and ^{13}C NMR.
10. ^1H NMR (300 MHz, CDCl3) δppm: 3.26 (2H, t, J12-11 = 7.40 Hz, H12); 1.70 (6H, s, H13); 1.65 (2H, m, H11); 1.36 (2H, m, H10); 1.23 (16H, br. s, H2-9); 0.86 (3H, t, J1-2 = 6.5 Hz, H1).

    ^{13}C NMR (500 MHz, CDCl3) δppm: 220.19 (C13); 177.76 (C16); 54.91 (C14); 36.46 (C12); 31.31, 29.02, 28.95, 28.84, 28.74, 28.50, 28.36, 22.08 (C2-9); 28.59 (C10); 27.19 (C11); 24.60 (C15); 13.52 (C1).

    IR ν: 2921 (CH2); 1714 (C=O); 1070 (S–(C=S)–S) cm^{-1}.

3.2 Preparation of 2-(Dodecylthio-carbonothioylthio)-2-methylpropionic Acid Pentafluorophenyl Ester (DMP-PFP)

1. Add 0.5 g (1.37 mmol) of DMP (*see* Subheading 3.1 for preparation), 0.39 g (2.05 mmol) of *N'*-(3-Dimethylaminopropyl)-*N'*-ethylcarbodiimide hydrochloride (EDC), and 0.25 g (2.05 mmol) of 4-(dimethylamino)pyridine (DMAP) in 40 mL of DCM and stir for 20 min at room temperature under N_2.
2. Add 0.76 g (4.24 mmol) of pentafluorophenol in 5 mL DCM and stir the reaction mixture was overnight at room temperature.

3. Wash the reaction solution successively with 3 M HCl (50 mL), 1 M $NaHCO_3$ (50 mL) and 0.5 M NaCl (50 mL).
4. Dry the reaction solution over $MgSO_4$, filter and then concentrate under reduced pressure, to afford a yellow oil that crystallises. No further purification is required.
5. Characterize the product using IR, ^1H, and ^{13}C NMR.

^1H NMR (300 MHz, CDCl3) δppm: 3.29 (2H, t, J12-11 = 7.3 Hz, H12); 1.84 (6H, s, H13); 1.67 (2H, q, J = 7 Hz, H11); 1.37 (2H, m, H10); 1.23 (16H, br. s, H2-9); 0.86 (3H, t, J1-2 = 6.5 Hz, H1).

^{13}C NMR (500 MHz, CDCl3) δppm: 220.19 (C13); 169.77 (C16); 54.91 (C14); 36.46 (C12); 31.31, 29.02, 28.95, 28.84, 28.74, 28.50, 28.36, 22.08 (C2-9); 28.59 (C10); 27.19 (C11); 24.60 (C15); 13.52 (C1).

IR ν: 2921 (CH2); 1779 (C6F5C=O); 1070 (S–(C=S)–S) cm^{-1}.

3.3 Polymerization of N-hydroxyethyl Acrylamide (HEA) (DP20) Using DMP-PFP

1. Dissolve 0.5 g (4.34 mmol) of *N*-hydroxyethyl acrylamide (HEA), 0.115 g (0.22 mmol) of PFP-DMP, 0.0122 g (0.043 mmol) of 4,4′-Azobis(4-cyanovaleric acid) (ACVA) in 50:50 toluene:methanol (4 mL).
2. Add mesitylene (150 μL) as an internal reference (*see* **Note 3**).
3. Degas under N_2 for 30 min and then stir at 70 °C for 90 min.
4. Rapidly cool in liquid nitrogen and precipitate into cold diethyl ether (45 mL). Reprecipitate into cold diethyl ether (45 mL) from methanol (3 mL) twice, to yield a yellow polymer product. Dry under vacuum.
5. Characterize the product using H^1 NMR, IR, and size-exclusion chromatography (SEC) (*see* **Note 4**).
6. 90 % conversion by NMR, M_n (Theoretical) = 2800 g/mol Mn (SEC) = 4100 g/mol Mn/Mw (SEC) = 1.14.

3.4 Functionalization of Polymer End Group

3.4.1 Method 1: Functionalization with Hexosamines

1. Dissolve 50 mg (0.018 mmol) of PFP-pHEA and hexosamine (15 mg, 0.088 mmol) in 5 mL of DMF (with TEA (0.05 M).
2. Stir at 50 °C for 16 h.
3. Precipitate into cold diethyl ether (45 mL) from methanol (3 mL) three times and dry under vacuum.
4. IR analysis indicates loss of C=O stretch corresponding to the PFP ester (*see* **Note 5**).

3.4.2 Method 2: Functionalization with Reducing Sugars

1. Dissolve 500 mg (0.18 mmol) of PFP-pHEA, 50 μL hydrazine (1.5 mmol) in 5 mL DMF.
2. Stir at 50 °C for 16 h.
3. Precipitate into cold diethyl ether (45 mL) from methanol (3 mL) three times and dried under vacuum.

4. IR indicates loss of C=O stretch corresponding to the PFP ester (*see* **Note 5**).
5. Add 10 mg (0.0043 mmol) of hydrazide-pHEA and reducing hexose (3 mg) in 1 mL of 100 mM acetate containing 1 mM aniline.
6. Leave it at 50 °C overnight.
7. Employ it immediately for gold nanoparticle functionalization.

3.5 Functionalization of Gold Nanoparticles

1. Add 1 mg of polymer in 100 μL water to 1 mL of 60 nm particles ($OD_{600}=1$, 0.288 mM Au) (*see* **Notes 6** and **7**).
2. Leave it for 30 min at room temperature.
3. Centrifuge the particle solution at 4600×*g*, and remove the supernatant along with any unattached polymer.
4. Resuspend the particles in water.
8. Characterize by DLS and NaCl titration (*see* **Notes 8** and **9**).
9. Average diameter increase of 5 nm determined by DLS (from 60 to 65 nm).

3.6 Lectin-Induced Aggregation

1. A panel of six lectins (Con A, RCA_{120}, SBA, UEA, PNA, and WGA) are used from Lectin Kit 1.
2. Make a stock solution of each lectin (0.1 mg/mL) in 10 mM HEPES buffer with 0.05 M NaCl, 0.1 mM $CaCl_2$, and 0.01 mM $MnCl_2$.
3. Make 25 μL serial dilutions (from 0.1 to 0 mg/mL) in the same HEPES buffer in a low volume 96-well micro-titer plate.
4. Add 25 μL of glycoAuNP to each well.

3.7 Data Collection and Analysis

3.7.1 Method 1: UV/Vis Spectrometry

1. 96-Well plates are prepared as above in Subheading 3.6.
2. Measure absorbance at 450 nm and 700 nm every minute for 30 min at 37 °C using a BioTek Synergy HT microplate reader.
3. After 30 min, record an absorbance spectrum from 450 to 700 nm with 10 nm intervals (1 nm intervals can be used for greater precision).
4. Normalize absorbance at 700 nm to that of reference wavelength 400 nm. Plot normalized 700 nm values on a logarithmic scale and fit to a Hill function. Determine K_d apparent value from this (Fig. 4) (*see* **Note 10**).

3.7.2 Method 2: Pixel Intensity Using Image Analysis

1. An image of the 96-well plate prepared in Subheading 3.6 is taken using a flatbed scanner after 30 min incubation at 37 °C.
2. A region of interest is drawn around every well using the open-source image analysis package Image J (version 1.46a).
3. The color (RGB) image is then converted into a hue, saturation, and brightness (HSB) stack of images and the saturation image used.

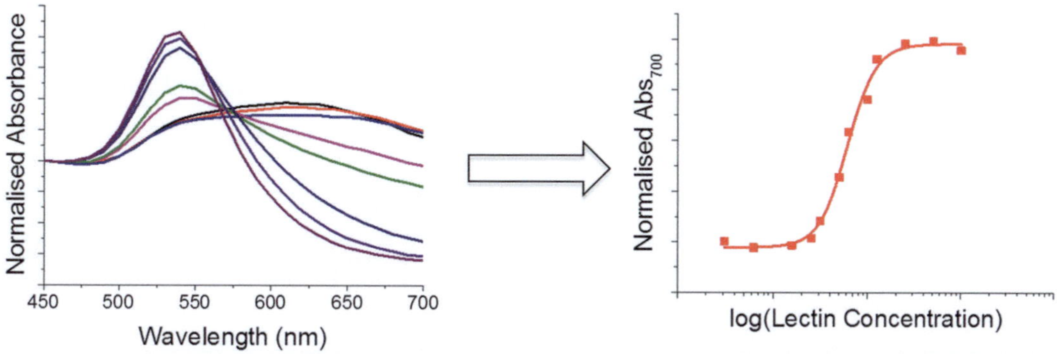

Fig. 4 Example data analysis: UV–Vis spectra of fucose-AuNP upon addition of increasing concentrations of UEA following 30 min of incubation. Binding isotherms (at 37 °C) of fucose-AuNPs with UEA from plotting normalized absorbance at 700 nm as a function of concentration

4. The regions of interest drawn on the original image are added to the saturation image using the ROI manager and average pixel intensity in each region of interest is measured using an inbuilt function in Image J.

4 Notes

1. Ca^{2+} and Mn^{2+} in HEPES buffer required for correct function of lectins. Do not use PBS as it binds Ca^{2+} and calcium phosphate precipitates.
2. Best results are obtained using fresh lectins. Lectins should be stored in buffer at 4–8 °C for no longer than 1 week.
3. ^{1}H NMR should be used obtain conversion. 25 μL of reaction mixture should be taken prior to reaction and ^{1}H NMR spectrum obtained, in $CDCl_3$. 25 μL of reaction mixture should be taken after reaction completion and ^{1}H NMR spectrum obtained in MeOD. Compare vinyl peaks of each sample to the internal standard mesitylene ($\delta = 6.8$ ppm).
4. SEC should be carried out in DMF. Note that the calibration standards do not give absolute M_n values. End-group analysis by ^{1}H NMR should be used to get a good estimate of DP.
5. Conversion of PFP ester C=O to amide C=O can be seen by disappearance of peak at around 1750 cm^{-1} or by ^{19}F NMR.
6. Commercial AuNPs should be checked before use for size discrepancies and dispersity to ensure they are in fact 60 nm as it is important for stability and readout speed [4]. Size characterization can be carried out using the SPR peak location [12], DLS and transmission electron microscopy (TEM).
7. AuNPs can be synthesized by citrate reduction method of Turkevich et al. [1]; however, commercial products will have lower dispersity [13].

8. Successful coating can be determined by a simple NaCl aggregation test. Coated particles will be stable beyond 0.1 M NaCl whereas uncoated citrate stabilized particles will aggregate.
9. DLS will confirm size increase due to coating and X-ray photoelectron spectroscopy (XPS) will give confirmation of surface functionalisation. TEM will not confirm coating.
10. To determine a comparable K_d apparent, the sigmoidal curves require an obvious start and end plateau as shown in Fig. 2.

Acknowledgements

Equipment used was supported by the Innovative Uses for Advanced Materials in the Modern World (AM2), with support from Advantage West Midlands (AWM) and part funded by the European Regional Development Fund (ERDF). MIG was a Birmingham Science City Interdisciplinary Research Fellow funded by the Higher Education Funding Council for England (HEFCE). SJR acknowledges the EPSRC funded MOAC doctoral training centre for a studentship. CIB acknowledges the BBSRC for a studentship.

References

1. Schofield CL, Mukhopadhyay B, Hardy SM, McDonnell MB, Field RA, Russell DA (2008) Colorimetric detection of *Ricinus communis* Agglutinin 120 using optimally presented carbohydrate-stabilised gold nanoparticles. Analyst 133:626–634
2. Lin CC, Yeh YC, Yang CY, Chen GF, Chen YC, Wu YC, Chen CC (2003) Quantitative analysis of multivalent interactions of carbohydrate-encapsulated gold nanoparticles with concanavalin A. Chem Commun 2920–2921
3. Jayawardena HSN, Wang X, Yan M (2013) Classification of lectins by pattern recognition using glyconanoparticles. Anal Chem 85:10277–10281
4. Richards S-J, Gibson MI (2014) Optimization of the polymer coating for glycosylated gold nanoparticle biosensors to ensure stability and rapid optical readouts. ACS Macro Lett 3:1004–1008
5. Lin CC, Yeh YC, Yang CY, Chen CL, Chen GF, Chen CC, Wu YC (2002) Selective binding of mannose-encapsulated gold nanoparticles to type 1 Pili in *Escherichia coli*. J Am Chem Soc 124:3508–3509
6. Richards S-J, Fullam E, Besra GS, Gibson MI (2014) Discrimination between bacterial phenotypes using glyco-nanoparticles and the impact of polymer coating on detection readouts. J Mater Chem B 2: 1490–1498
7. Marin MJ, Rashid A, Rejzek M, Fairhurst SA, Wharton SA, Martin SR, McCauley JW, Wileman T, Field RA, Russell DA (2013) Glyconanoparticles for the plasmonic detection and discrimination between human and avian influenza virus. Org Biomol Chem 11:7101–7107
8. Gibson MI, Paripovic D, Klok H-A (2010) Size-dependent LCST transitions of polymer-coated gold nanoparticles: cooperative aggregation and surface assembly. Adv Mater 22:4721–4725
9. Ieong NS, Brebis K, Daniel LE, O'Reilly RK, Gibson MI (2011) The critical importance of size on thermoresponsive nanoparticle transition temperatures: gold and micelle-based polymer nanoparticles. Chem Commun 47:11627–11629
10. Biggs CI, Edmondson S, Gibson MI (2015) Thiol-ene immobilisation of carbohydrates

onto glass slides as a simple alternative to gold-thiol monolayers, amines or lipid binding. Biomater Sci 3:175–181

11. Godula K, Bertozzi CR (2010) Synthesis of glycopolymers for microarray applications via ligation of reducing sugars to a poly(acryloyl hydrazide) scaffold. J Am Chem Soc 132: 9963–9965
12. Haiss W, Thanh NTK, Aveyard J, Fernig DG (2007) Determination of size and concentration of gold nanoparticles from UV–vis spectra. Anal Chem 79:4215–4221
13. Turkevich J, Stevenson PC, Hillier J (1951) A study of the nucleation and growth processes in the synthesis of colloidal gold. Discuss Faraday Soc 11:55–75

Part III

Surface Immobilized Glycopolymers

Chapter 15

Modulation of Multivalent Protein Binding on Surfaces by Glycopolymer Brush Chemistry

Kai Yu, A. Louise Creagh, Charles A. Haynes, and Jayachandran N. Kizhakkedathu

Abstract

The presentation of carbohydrates on an array can provide a means to model (mimic) oligosaccharides found on cell surfaces. Tuning the structural features of such carbohydrate arrays can therefore be used to help to elucidate the molecular mechanisms of protein-carbohydrate recognition on cell surfaces. Here we present a strategy to directly correlate the molecular and structural features of ligands presented on a surface with the kinetics and affinity of carbohydrate–lectin binding. The Surface Plasmon Resonance (SPR) spectroscopy analysis identified that by varying the spatial distribution (3D organization) of carbohydrate ligands within the surface grafted polymer layer, the mode of binding changed from multivalent to monovalent: a near 1000-fold change in the equilibrium association constant was achieved. The rupture forces measured by atomic force microscopy (AFM) force spectroscopy also indicated that the mode of binding between lectin and carbohydrate ligands can be modulated by the organization of carbohydrate ligands within the glycopolymer brushes.

Key words Glycopolymers, Polymer brush, Lectin, Carbohydrate, Binding kinetics

1 Introduction

The interaction of lectins and other carbohydrate-binding proteins with glycoproteins, glycolipids or polysaccharides displayed on a cell surface mediate a variety of biological processes such as cell signaling, cell adhesion, fertilization, and inflammatory responses [1–3]. Since the affinity of single carbohydrate residue and a protein is usually weak, multivalent interactions are often used by the nature by displaying multiple ligands or by the use of multivalent proteins [4–8]. Multivalency is often achieved through chelating effect [6–8] and proximity/statistical effect [9]. The increase in the affinity is thought to be dependent on the structure of the lectin [10], the density of the carbohydrate ligands and the manner in which they are presented on the cell surface [11–13]. An improved understanding of the contribution of the molecular

Xue-Long Sun (ed.), *Macro-Glycoligands: Methods and Protocols*, Methods in Molecular Biology, vol. 1367,
DOI 10.1007/978-1-4939-3130-9_15, © Springer Science+Business Media New York 2016

and structural features of ligands presented on a surface to the mechanisms of carbohydrate–protein recognition would therefore be of great use, in part by helping to resolve their roles in various biological processes.

While improvements in the affinity of multivalent lectins for carbohydrates presenting surfaces have been achieved, our fundamental understanding of the reaction mechanism connecting binding affinity/avidity and surface structure remains poor. This limitation could be addressed through the development of carbohydrate arrays where the binding interactions can be tuned from monovalent to multivalent using chemistry that allows for precise control of the presentation of the ligands on a surface. The implementation of such a technology would permit direct correlation of binding affinity/avidity with engineered changes in the surface structure, and thereby aid the design and development of potent lectin inhibitors and biological effectors.

In this study, we therefore report on a strategy to modulate multivalent interactions at the surfaces by changing the spatial arrangement of carbohydrate residues (mannose and galactose) within end-grafted glycopolymer brushes [14]. Glycopolymer brushes with pendent sugar residues (Fig. 1) are synthesized with tight control of grafting density, degree of polymerization, and carbohydrate composition; all these parameters are independently controlled. These systems offer an opportunity to mimic the cell surface glycocalyx, as sugar units along each grafted polymer chain is presented in a manner similar to proteoglycans presented on the cell surface. A systematic investigation on the interaction of *Concanavalin A* (*Con A*), a bivalent lectin, with glycopolymer brushes was conducted using SPR to explore the influence of molecular and structural features of the surface on binding characteristics. Binding equilibrium and kinetic data collected by SPR are reported and are used to determine equilibrium, forward and reverse rate constants for each possible association/dissociation

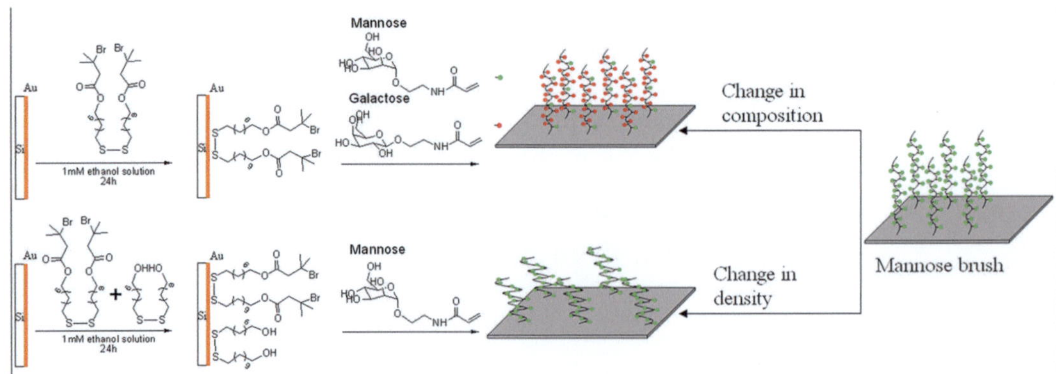

Fig. 1 Synthesis of glycopolymer brush with different composition and grafting density

reaction that can occur between bivalent Con A and the grafted brush layer displaying the binding partner, mannose. The intermolecular unbinding force between Con A and carbohydrate units on the surface is also investigated by AFM force spectroscopy.

2 Materials

2.1 Chemicals

1. 2′-acrylamidoethyl-α-D-mannopyranoside (AAEM, mannose monomer) and 2′-acrylamidoethyl-β-D-galactopyranoside (AAEGal, galactose monomer) are synthesized by a procedure as reported previously [15].
2. Copper (I) chloride (99 %, Sigma-Aldrich) is purified by washing with acetic acid, methanol and dried under vacuum.
3. Copper (II) chloride (99 %, Sigma-Aldrich).
4. Tris (2-(dimethylamino)ethyl)amine (Me_6TREN) (97 %, Sigma-Aldrich).
5. Methyl 2-chloropropionate (97 %, Sigma-Aldrich).
6. 2-bromo-2-isobutyryl bromide (98 %, Sigma-Aldrich).
7. Manganese (II) chloride (99 %, Sigma-Aldrich).
8. Calcium chloride (99 %, Sigma-Aldrich).
9. D-(+)-mannose (99 %, Sigma-Aldrich).
10. Phosphoric acid solution (85 wt% in H_2O, Sigma-Aldrich).
11. Triethylamine (Et_3N) (99 %, Sigma-Aldrich).
12. Ethanolamine hydrochloride (99 %, Sigma-Aldrich).
13. DMSO (99 %, Sigma-Aldrich).
14. NHS-PEG6000-NHS (Rapp Polymere GmbH, Germany).
15. Surface ATRP initiator $(BrC\text{-}(CH_3)_2COO(CH_2)_{11}S)_2$ (SAM-Br) and $(HO(CH_2)_{11}S)_2$ (SAM-OH) are synthesized by using a similar procedure reported in the literature [16].

2.2 Other Commercial Regents

1. SPR chip: Bare gold chip (BIAcore SIA kit Au, Catalog No. BR-1004-05) is purchased from GE Healthcare.
2. SPR Maintenance kit (BR-1006-66) is purchased from GE Healthcare.
3. AFM probes with spring constant about 0.06 N/m are purchased from Veeco, US.

2.3 Buffering Solutions

1. 150 mM phosphate buffered solution, pH 4.8, 137 mM NaCl, 10 mM phosphate, 2.7 mM KCl, containing 1 mM $CaCl_2$ and 1 mM $MnCl_2$.
2. 25 mM phosphate buffered solution, pH 4.8.
3. 25 mM phosphate buffered solution, pH 7.4.

4. 5 M mannose buffer solution.
5. 100 mM phosphoric acid solution.

2.4 Proteins

Concanavalin A from *Canavalia ensiformis* (Jack bean) Type IV, lyophilized powder (Sigma).

3 Methods

3.1 Preparation of Glycopolymer Brushes on Au Chips

The glycopolymer brushes with different composition and grafting density were grown on initiator modified SPR bare Au chip by surface-initiated atom transfer radical polymerization through changing the monomer composition and initiator density. End-grafted galactose-based copolymers were prepared with increasing amounts of mannose to investigate the influence of various characteristics of the glyco-structures on carbohydrate binding of the lectin, Con A. Galactose is a neutral sugar which has no specific binding interaction with Con A and was therefore be used to dilute the ligand concentration within the brush without affecting the properties of brush structure such as hydration and steric factors. In addition, mannose homopolymer brushes with different grafting density were prepared to investigate whether steric factors influence the binding characteristics.

SPR Analyses: SPR measurements were performed on a BIAcore 3000 (BIAcore, Uppsala, Sweden) operated using the BIAcore control software. The flow rate of analyte solution and 150 mM PBS buffer (pH 4.8, containing 1 mM Ca^{2+}, 1 mM Mn^{2+}) through the flow cells was set at 30 μL/min. A series of 750 μL Con A solutions were injected and allowed to flow through the channels. The concentration of each injection varied from 0.05 to 150 μM. There was no regeneration between consecutive injections (Fig. 2a). The pAAEGal brush modified gold chip was set as the reference surface since there was no specific interaction observed between Con A and pAAEGal brushes. The response difference between mannose and galactose polymer brush modified surfaces was taken as the response due to the specific interaction between Con A and mannose containing brush. The equilibrium response was determined from duplicate measurements and plotted against the analyte concentration, and fitted to a model with Hill slope ([17], Eq. 1) by using the GraphPad Prism 5 to determine the association constant (Table 1) [14].

$$\mathrm{RU} = \mathrm{RU}_{\max} \frac{C^h}{K_\mathrm{D} + C^h} \tag{1}$$

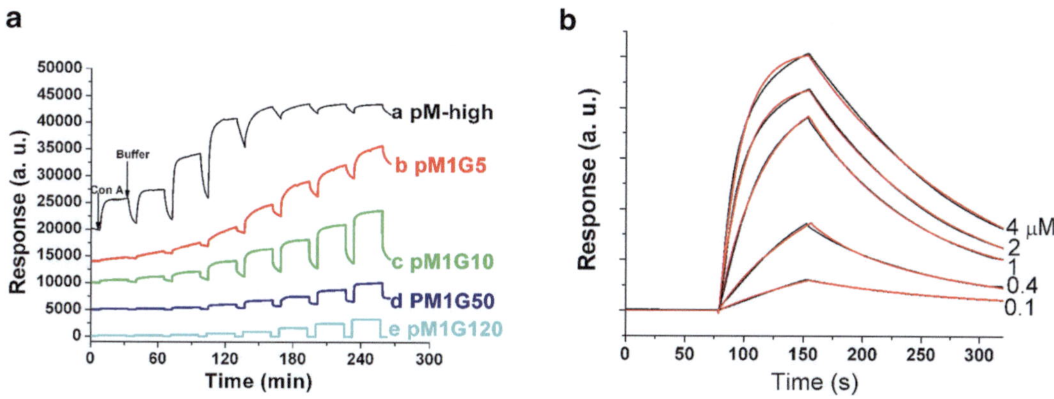

Fig. 2 (**a**) SPR sensorgram for the interaction of Con A with glycopolymer layers with different mannose contents: (a) 100 %; (b) 16.7 %; (c) 9 %; (d) 2 %; and (e) 0.8 %. For the copolymer layers composing of 100 and 16.7 % mannose, eight different concentrations (in the sequence of injection, 0.1, 0.2, 0.4, 1, 4, 10, 20, 50 μM) of Con A was injected and flowed through. For the copolymer layers composing of 9 % and 2 % mannose, eight different concentrations (1, 2, 4, 10, 20, 32, 64, 108 μM) of Con A was injected and flowed through the channel. For the copolymer layer composing of 0.8 % mannose, eight different concentration (1, 4, 10, 20, 32, 64, 108, 150 μM) of Con A was injected and flowed through the channel. (**b**) Kinetics analyses of Con A binding to glycopolymer layers. Shown are experimental data (*black curves*) and the derivation of the fitted sensorgram (*red lines*) by bivalent model fitting of Con A binding to glycopolymer layers with mannose content 100 %. Reproduced with permission from Yu K, Creagh AL, Haynes CA, Kizhakkedathu JN (2013) *Anal Chem* 85, 7786–7793. Copyright (2013) American Chemistry Society

where K_D (=$1/K_A$) is the equilibrium dissociation constant, h is the Hill coefficient, RU is the steady-state SPR response to a solution of Con A of concentration C, and RU_{max} is the maximal SPR response brought by binding extrapolated to the saturation concentration of lectin.

For the calculation of the association rate constant and dissociation rate constant, a conventional regeneration protocol was used. After each injection of Con A solution, the chip was regenerated (Fig. 2b). The association and dissociation rate constants were derived by fitting the data to a bivalent analyte model using the BIAevaluation Software, version 4.1. Bulk refractive index effect brought by variation of analyte concentration was fixed during the fitting. The bulk effect was measured by allowing analyte solution flow through the reference sample-galactose brush modified chip. The dissociation rate constant for the interaction between Con A and brush presenting 0.8 % mannose was obtained by fitting the data to one phase exponential decay with using Origin 7.0 software [14].

3.2 AFM Force Spectroscopy Studies of Lectin Binding to Glycopolymer Brushes

Glycopolymer brushes having compositions and structures similar to those employed in the SPR analyses were utilized. AFM tips were functionalized with Con A using PEG as a spacer. The approach and retraction force curves for a Con A-coupled AFM tip to either an all mannose brush or an all galactose brush were

Table 1
Equilibrium association constant (K_A) and Hill coefficient (h) data for binding of Con A to various end-grafted glycopolymers

Samples	Mannose composition (%)	Grafting density, σ, (chains/nm^2)	Association constant (K_A) from Eq. 1 (M^{-1})	Association constant (K_A) from kinetic study (M^{-1})	h
pM-low	100	0.006	$(7.4 \pm 1.3) \times 10^5$	N/D	0.55 ± 0.14
pM-med	100	0.02	$(1.5 \pm 0.6) \times 10^6$	N/D	0.71 ± 0.09
pM-high	100	0.10	$(3.1 \pm 0.5) \times 10^6$	5.5×10^6	0.69 ± 0.12
pM1G5	16.7	0.09	$(1.2 \pm 0.5) \times 10^5$	6.2×10^5	0.65 ± 0.06
pM1G10	9	0.08	$(3.1 \pm 0.2) \times 10^4$	N/D	0.80 ± 0.05
pM1G20	4.8	0.09	$(2.6 \pm 0.3) \times 10^4$	5.2×10^4	0.83 ± 0.04
pM1G50	2	0.09	$(8.4 \pm 0.3) \times 10^3$	9.2×10^3	0.96 ± 0.06
pM1G120	0.8	0.09	$(3.8 \pm 0.6) \times 10^3$	4.2×10^3	0.99 ± 0.10

Reprinted with permission from Yu K, Creagh AL, Haynes CA, Kizhakkedathu JN (2013) *Anal Chem* 85, 7786–7793. Copyright (2013) American Chemistry Society

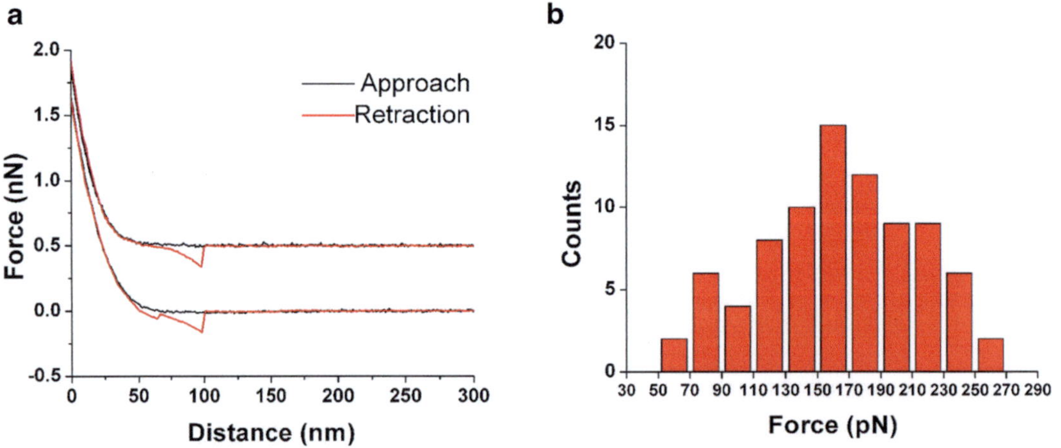

Fig. 3 (**a**) Representative approach (*black line*) and retraction (*red line*) force curves for the interaction of mannose polymer layer ($\sigma = 0.1$ chains/nm^2 and height = 59 nm) with Con A modified tip. (**b**) Probability distribution histograms of the maximum unbinding force during retracting Con A modified tip from mannose polymer layer. Reprinted with permission from Yu K, Creagh AL, Haynes CA, Kizhakkedathu JN (2013) *Anal Chem* 85, 7786–7793. Copyright (2013) American Chemistry Society

recorded (Fig. 3). On tip approach, the onset of the region of constant compliance was used to determine the zero distance, and on retraction the region in which the force was unchanged was used to determine the zero force. The rate of tip-sample approach or retraction was set as 0.5 μm/s. The raw AFM force data (cantilever deflection vs. displacement data) were converted into force vs. separation following the principle of Ducker et al. by using custom Matlab v.5.3 (Math Works, Natick, MA) software [18]. The software converts the cantilever deflection vs. linear voltage displacement transformer signal into restoring force vs. tip-substrate separation using user input trigger and spring constant values. We followed our published protocol for the calculations of the adhesive force [19].

3.3 Preparation of End-Grafted Glycopolymer on SPR Chip

3.3.1 Preparation of Initiator Modified Gold Chip

1. Clean Gold coated chips first by soaking in chromic acid for 1 h. Then wash the chips with water and dry under a stream of argon.
2. Deposit self-assembled monolayers onto the gold by immersing the cleaned gold chips in 1 mM SAM-Br or mixtures of SAM-Br and SAM-OH overnight at room temperature (22 °C).
3. Wash the initiator modified chips thoroughly with ethanol and THF and dry in an argon flow.

3.3.2 Synthesis of pAAEM Layers on the Initiator Modified Au Chip

1. Add copper (II) chloride ($CuCl_2$, 1.35 mg, 0.01 mmol), copper (I) chloride (CuCl, 8 mg, 0.08 mmol), Me_6TREN (52 μL, 0.18 mmol) successively into a glass tube followed by Milli-Q water (12 mL).
2. Degas the catalyst solution with three freeze pump–thaw cycles.
3. Transfer the solution into a glove box.
4. Draw aliquots of the catalyst solution (1.5 mL) and add into the vials which contained 2′-acrylamidoethyl-α-D-mannopyranoside (75 mg). After the monomer was completely dissolved, immerse the ATRP initiator modified gold chip fully into the solution.
5. Add 20 μL of a solution of methyl 2-chloropropionate in methanol (250 μL in 5 mL methanol) to the reaction solution.
6. Allow the surface-initiated polymerization to proceed at RT (22 °C) for 24 h.
7. Rinse the substrates with water thoroughly, followed by drying under a flow of argon gas.
8. The solution was dialyzed against deionized water, and then freeze-dry. *See* **Note 1** for overall.

3.4 SPR Measurements

3.4.1 SPR Equilibrium Studies

1. Assemble the brush modified chip onto the chip holder and dock into the SPR instrument.
2. Normalize signal response by using BIAnormalzing solution provided in the maintenance kit.
3. Inject buffer (150 mM phosphate buffered solution, pH 4.8, with 1 mM $CaCl_2$ and 1 mM $MnCl_2$) and allow to flow through

the channels of SPR chip until the baseline stabilized with changes less than 1RU/min.

4. Inject a series of 750 μL Con A solutions (0.05–150 μM) at a flow rate of 30 μL/min and allowed to flow through the channels without regeneration between consecutive injections.
5. Undock the chip and replace with maintenance chip.
6. Desorb procedure is done using the desorb solution 1 and 2 providing in the maintaining kit. *See* **Note 2** for overall.

3.4.2 SPR Kinetic Studies

The procedure is similar to the SPR equilibrium study except that Con A solution is injected and allowed to flow through the channels. After each injection, the chip is regenerated by 300 μL 5 M mannose buffer solution twice and 10 μL 0.1 M phosphoric acid. *See* **Note 3** for overall.

3.5 AFM Force Spectroscopy

3.5.1 AFM Tip Modification

1. Clean AFM tips by O_2 plasma at 50 W for 60 s.
2. Immerse the AFM tips overnight in a solution of ethanolamine hydrochloride in DMSO (5.5 g in 10 mL) to generate amino groups on the tip surface.
3. Rinse the tips in chloroform and incubate in a chloroform solution containing NHS-PEG6000-NHS (10 mg/mL) containing 0.5 % Et_3N.
4. Rinse the tips with chloroform, dry with argon and immerse in 25 mM phosphate buffer (pH 7.4) containing Con A (2 mg/mL). After 2 h, remove the tips from incubation and wash it briefly in 25 mM phosphate buffer (pH 7.4) and then with 25 mM phosphate buffer (pH 4.8).

3.5.2 AFM Force Measurements

1. Mount the brush modified substrate on the stage using an AFM steel punk and sticky tab.
2. Attach the Con A modified tip to the cantilever holder on a wet cell.
3. Injected buffer solution (150 mM phosphate buffered solution, pH 4.8, with 1 mM $CaCl_2$ and 1 mM $MnCl_2$) through the hole of the wet cell to make the cantilever and samples fully wetted.
4. After laser alignment and photodiode alignment, set scanning parameters and engaging the microscope and bring the tips in contact with the substrate for 10 min.
5. Collect force curves with ramp size of 500 nm and rate of tip-sample approach or retraction was set as 0.5 μm/s. *See* **Note 4** for overall.

4 Notes

1. Poly(2′-acrylamidoethyl-β-D-galactopyranoside) (pAAEGal) layers and copolymers of 2′-acrylamidoethyl-α-D-mannopyranoside and 2′-acrylamidoethyl-β-D-galactopyranoside P(AAEM-*co*-AAEGal) layers with different molar feed ratios of AAEM to AAEGal (1:5, 1;10, 1:20, 1:50, and 1: 120) are synthesized using a similar procedure. All the layer structures are characterized in terms of thickness, grafting density, and molecular weight (Table 2).

Table 2
Characteristics of pAAEM (pM, M-mannose) layers and p(AAEM-*co*-AAEGal) (pM_XG_Y, M-mannose, G-galactose) layers

Samples	Feed ratio of AAEM (mannose) to AAEGal (galactose)	Mannose composition (%)	Dry thickness (nm)	Grafting density, σ^a (chains/ nm^2)	Molecular weight of free polymer (Mn) (Mw/Mn)	Distance between the chains, d^b (nm)	Equilibrium thickness, Le^c (nm)
pM-high	100, 0	100	19.6	0.10	123 000 (1.3)	3.2	59 ± 3.6
pM-med	100, 0	100	4.6	0.02	130 000 (1.3)	7.1	15.5 ± 0.7
pM-low	100, 0	100	1.3	0.006	130 000 (1.3)	12.9	N/D
pM1G5	1, 5	16.7	20	0.09	131 000 (1.3)	3.3	74 ± 5.7
pM1G10	1, 10	9	16.3	0.08	122 000(1.4)	3.5	55 ± 3.2
pM1G20	1, 20	4.8	22	0.09	141 000 (1.4)	3.3	67 ± 4.7
pM1G50	1, 50	2	23	0.09	148 000 (1.3)	3.3	71 ± 6.3
pM1G120	1, 120	0.8	17	0.09	118 000 (1.4)	3.3	58 ± 5.1
pG	0, 100	0	24.4	0.09	158 000 (1.3)	3.3	70 ± 2.7

[a]The grafting density (σ) for glycopolymer layers was estimated by using the equation, $\sigma = (h\rho N_A)/Mn$, where Mn is the molecular weight of free polymer in the solution, N_A is the Avogado's number, h is the polymer layer thickness measured by elliposometer, ρ is the density of glycopolymer (we assumed the density of glycopolymer is equal to 1 g/cm^3)

[b]
$$d = \frac{1}{\sqrt{\sigma}} \text{nm}$$

[c]Equilibrium thickness (Le) was determined by AFM as the critical distance from the substrate surface beyond which no repulsive force was detectable. *N/D* not determined. Reprinted with permission from Yu K, Creagh AL, Haynes CA, Kizhakkedathu JN (2013) *Anal Chem* 85, 7786–7793. Copyright (2013) American Chemistry Society

2. Procedure for the calculation of binding constant was described in the method section SPR analyses.
3. The bulk effect was measured by allowing analyte solution flow through the reference sample-galactose layer modified chip. Procedure for calculation of the association and dissociation rate constants was described in the previous method section on SPR analyses.
4. The conversion of raw AFM force data to force vs. separation and the calculations of the adhesive force are given in the section "AFM force spectroscopy studies of lectin binding to glycopolymer brushes." The spring constant for the AFM cantilever was measured using thermal equipartition theorem [20]. The force curves collected on the galactose brush was used to observe the no specific interaction.

Acknowledgments

The authors acknowledge the funding provided by the NSERC and CIHR. The LMB Macromolecular and Biothermodynamics Hubs at the UBC were funded by Canada Foundation for Innovation and Michael Smith Foundation of Health Research. K.Y. is a recipient of a CIHR/Canadian Blood Services (CBS) postdoctoral fellowship in Transfusion Science. J.N.K. is a recipient of a Michael Smith Foundation of Health Research Career Investigator Award.

References

1. Sharon N, Lis H (2004) History of lectins, from hemagglutinins to biological recognition molecules. Glycobiology 14:53R–62R
2. Bertozzi CR, Kiessling LL (2001) Chemical glycobiology. Science 291:2357–2364
3. Dube DH, Bertozzi CR (2005) Glycans in cancer and inflammation – potential for therapeutics and diagnostics. Nat Rev Drug Discov 4:477–488
4. Mandal DK, Kishore N, Brewer CF (1994) Thermodynamics of lectin-carbohydrate interactions. titration microcalorimetry measurements of the binding of N-linked carbohydrates and ovalbumin to concanavalin A. Biochemistry 33:1149–1156
5. Schwarz FP, Puri KD, Bhat RG, Surolia A (1993) Thermodynamics of monosaccharide binding to concanavalin A, pea (Pisum sativum) lectin, and lentil (Lens culinaris) lectin. J Biol Chem 268:7668–7677
6. Mammen M, Choi SK, Whitesides GM (1998) Polyvalent interactions in biological systems, implications for design and use of multivalent ligands and inhibitors. Angew Chem Int Ed 37:2755–2794
7. Lundquist JJ, Toone EJ (2002) The cluster glycoside effect. Chem Rev 102:555–578
8. Dam TK, Brewer CF (2008) Effects of clustered epitopes in multivalent ligand-receptor interactions. Biochemistry 47:8470–8476
9. Wolfenden ML, Cloninger MJ (2006) Carbohydrate-functionalized dendrimers to investigate the predictable tunability of multivalent interactions. Bioconjugate Chem 17:958–966
10. Ting SRS, Chen G, Stenzel MH (2010) Synthesis of glycopolymers and their multivalent recognitions with lectins. Polym Chem 1:1392–1412
11. Godula K, Bertozzi CR (2012) Density variant glycan microarray for evaluating cross-linking

of mucin-like glycoconjugates by lectins. J Am Chem Soc 124:15732–15742

12. Wilczewski M, Van der Heyden A, Renaudet O, Dumy P, Coche-Guerente L, Labbe P (2008) Promotion of sugar–lectin recognition through the multiple sugar presentation offered by regioselectively addressable functionalized templates (RAFT), a QCM-D and SPR study. Org Biomol Chem 6:1114–1122
13. Wang X, Ramstrom O, Yan MD (2010) Quantitative analysis of multivalent ligand presentation on gold glyconanoparticles and the impact on lectin binding. Anal Chem 82:9082–9089
14. Yu K, Creagh AL, Haynes CA, Kizhakkedathu JN (2013) Lectin interactions on surface-grafted glycostructures, influence of the spatial distribution of carbohydrates on the binding kinetics and rupture forces. Anal Chem 85:7786–7793
15. Yu K, Kizhakkedathu JN (2010) Synthesis of functional polymer brushes containing carbohydrate residues in the pyranose form and their specific and nonspecific interactions with proteins. Biomacromolecules 11:3073–3085
16. Shah RR, Merreceyes D, Husemann M, Rees I, Abbott NL, Hawker CJ, Hedrick JL (2000) Using atom transfer radical polymerization to amplify monolayers of initiators patterned by microcontact printing into polymer brushes for pattern transfer. Macromolecules 33: 597–605
17. Dam TK, Gerken TA, Brewer CF (2009) Thermodynamics of multivalent carbohydrate-lectin cross-linking interactions, importance of entropy in the bind and jump mechanism. Biochemistry 48:3822–3827
18. Ducker WA, Sendan TJ, Pashley RM (1992) Measurement of forces in liquids using a force microscope. Langmuir 8:1831–1835
19. Zou YQ, Rossi NAA, Kizhakkedathu JN, Brooks DE (2009) Barrier capacity of hydrophilic polymer brushes to prevent hydrophobic interactions, effect of graft density and hydrophilicity. Macromolecules 42:4817–4828
20. Wang M, Cao Y, Li HB (2006) The unfolding and folding dynamics of TNfnALL probed by single molecule force-ramp spectroscopy. Polymer 47:2548–2554

Chapter 16

Oriented Immobilized Sialyloligo-macroligand Microarray

Satya Nandana Narla and Xue-Long Sun

Abstract

Silaic acid is the most common terminal glycan on cell surface glycoproteins and glycolipids, involving in many biological processes. Studying interactions between multivalent sialic acid scaffolds and proteins binding to it will give incredible information in understanding the biological process the sialic acid is involved in. Here we describe chemoenzymatic synthesis of chain-end functionalized sialyllactose-containing glycopolymers with different linkages and their oriented immobilization for glycoarray and SPR-based glyco-biosensor applications.

Key words Sialic acid, Glycan array, Glycopolymer, Lectin, Hemagglutinin, SPR

1 Introduction

Sialic acids are a family of 9-carbon containing acidic monosaccharides, often terminate oligosaccharide structures of cell surface glycoconjugates such as glycoproteins and glycolipids, and are involved in many biological recognition systems. For example, sialic acids expressed on respiratory epithelial cell surface are involved in influenza virus infection in both virus hemagglutinin-based attaching [1] and neuraminidase-based detaching processes [2]. Influenza virus contains about 350–400 HA trimmers [3] and 50 NA tetramers [4] on its surface, which facilitate strong binding affinity through multivalent interactions with the sialic acid receptors on the host cell surface. Previous efforts using sialic acid-bearing macromolecules provided a proof of the concept of multivalent hemagglutinin inhibition [5]. However, the monosaccharide sialic acid cannot account for the molecular determinant of virus receptor-binding specificity in the context of the whole sialyloligosaccharide receptor. Recently, it has been known that the host cell binding specificity and affinity of influenza viruses depends not only on sialic acid but also the penultimate sugar and the linkage of sialic acid to the penultimate sugar on the receptor.

Xue-Long Sun (ed.), *Macro-Glycoligands: Methods and Protocols*, Methods in Molecular Biology, vol. 1367,
DOI 10.1007/978-1-4939-3130-9_16, © Springer Science+Business Media New York 2016

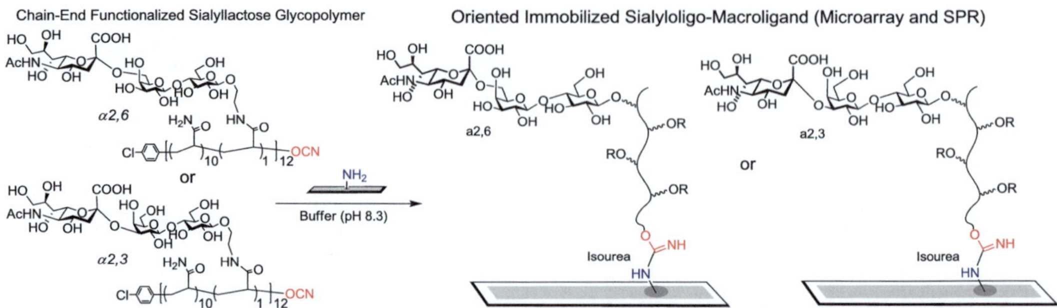

Fig. 1 Chemical structure of *O*-cyanate chain-end functionalized sialyllactose-containing glycopolymers and their orientally immobilized glyco-macroligand formation via isourea bond formation for microarray and SPR applications

Glycopolymers, namely polymers with carbohydrate pendent groups, have been extensively explored for different biological and biomedical applications [6, 7]. It is generally accepted that glycopolymers can mimic the functions of naturally occurring glycoconjugates [8, 9] and have been employed as cell-surface binding receptors [10]. Herein, we report a chemoenzymatic synthesis of *O*-cyanate chain-end functionalized sialyllactose-containing glycopolymers and their oriented sialyloligo-marcroligand formation for glycoarray and glyco-biosensor applications (Fig. 1). This oriented sialyloligo-macroligand platforms are expected to facilitate both affinity and specificity of protein binding and thus will provide a versatile tool for profiling glycan recognition for both basic biological research and practical applications.

2 Materials

2.1 Chemicals

1. 2-*N*-acryloyl-aminoethoxyl 4-*O*-(*β*-D-galactopyranosyl)-*β*-D-glucopyranoside) is synthesized as previously described [11] (*see* **Note 1**).
2. 4-Chloroaniline.
3. Sodium nitrite ($NaNO_2$).
4. Tetrafluoroboric acid (HBF_4) (48 wt%).
5. Acrylamide.
6. Sodium cyanate (NaOCN).
7. Sulfuric Acid (H_2SO_4).
8. Cystamine.
9. Hydrogen peroxide (H_2O_2).
10. Hydrochloric Acid (HCl).
11. Sodium hydroxide (NaOH).
12. Phenol.

13. Sialic acid (Rose Scientific Ltd).
14. α2,3-Sialyllactose (V-Labs).
15. α2,6-Sialyllactose (V-Labs).

2.2 Other Commercial Reagents and Solvents

1. Methanol.
2. Ethanol (Fisher scientific).
3. Sephadex G-25 (Sigma Aldrich).
4. Tetrahydrofuran (THF).
5. Amine functionalized glass slides (Xenopore, Co).
6. MicroCaster microarray tool (Whatman, spot size 500 μm diameter).

2.3 Buffers

1. 0.10 M $NaHCO_3$ buffer (pH 10.3).
2. 0.10 M PBS buffer (pH 7.4) containing NaCl (137 mM), KCl (2.7 mM), Na_2HPO_4 (10 mM), KH_2PO_4 (1.8 mM).
3. 0.10 M PBST buffer (pH 7.4) containing 0.02 % Tween 20.
4. 0.10 M HEPES (pH 7.4) containing Triton X-100 (0.5 %), $MgCl_2$ (5 mM), $MnCl_2$ (2 mM), $ZnCl_2$ (5×10^{-5} M).

2.4 Proteins and Enzymes

1. MAA-FITC (*Maackia amurensis*, 1.4×10^{-3} mM, Bioworld).
2. SNA-FITC (*Sambucus nigra*, 1.3×10^{-3} mM, Bioworld).
3. PNA-FITC (*Arachis hypogaea*, 1.6×10^{-3} mM, Sigma).
4. CMP-Neu5Ac.
5. Calf alkaline phosphatase.
6. Bovine serum albumin.
7. α2,3-Sialyltransferase.
8. α2,6-Sialyltransferase.
9. H5N1 HA (A/Anhui/1/2005(H5N1)) (Sino Biological, China).
10. H3N2 HA (A/Brisbane/10/2007(H3N2)) (Sino Biological, China).

3 Methods

The synthesis of end-functionalized glycopolymers still poses significant challenges, including a requirement for serial protection/deprotection steps or further polymer derivatization after initial synthesis. The *O*-cyanate chain-end functionalized glycopolymers were synthesized *via* our previously developed cyanoxyl free radical mediated copolymerization scheme in one-pot fashion (Fig. 2) [12]. Briefly, 4-chloroaniline was used as initiator for the

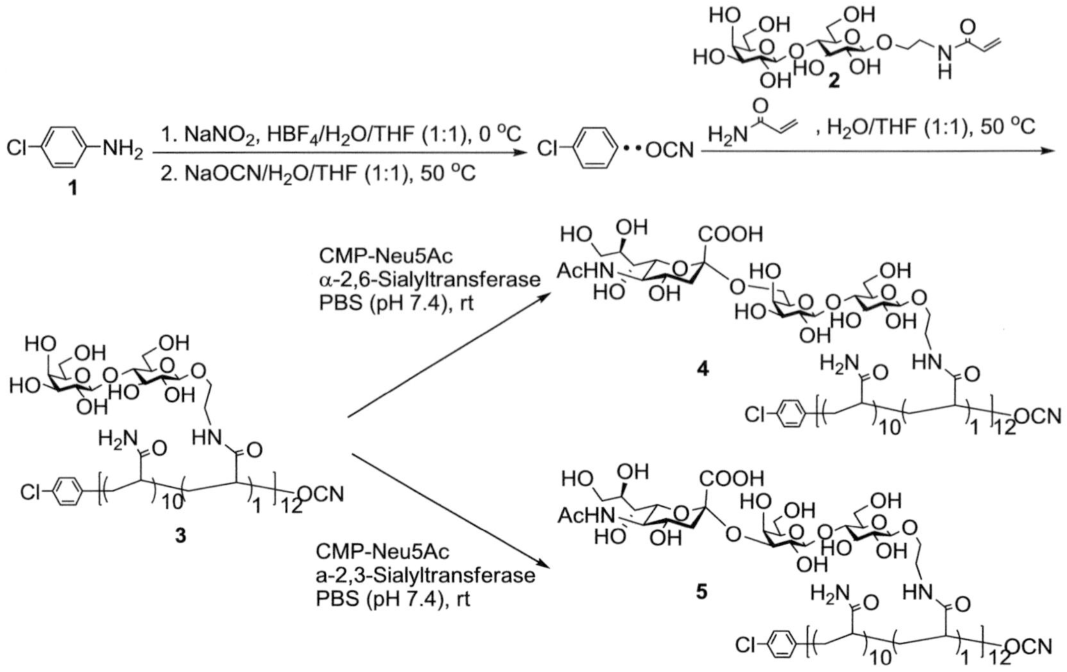

Fig. 2 Chemoenzymatic synthesis of *O*-cyanate chain-end functionalized sialyllactose-containing glycopolymers via cyanoxyl mediated free radical polymerization followed by enzymatic sialylation

copolymerization of acrylaminoethyl lactoside and acrylamide. Specifically, cyanoxyl radicals were generated by an electron-transfer reaction between cyanate anions from a sodium cyanate aqueous solution and aryl-diazonium salts prepared in situ through a diazotization reaction of arylamine in water. In addition to cyanoxyl persistent radicals, aryl-type active radicals were simultaneously produced, and only the latter species was capable of initiating chain growth. A series of *O*-cyanate chain-end functionalized glycopolymers of varying sugar density were prepared by altering glycomonomer (GM) and acrylamide (AA) in good conversion yield (60–70 %) and low polydispersity ($1.1 < Mw/Mn < 1.6$) as described previously [13].

The *O*-cyanate chain-end functionalized lactose-containing glycopolymer (**3**) synthesized was then subjected to enzymatic sialylation of the terminal galactose (Gal) pending on the polymer for synthesizing *O*-cyanate chain-end functionalized sialyllactose-containing glycopolymer (Fig. 2). The enzymatic approach for transferring sialic acid from CMP-Neu5Ac to the non-reducing end of the carbohydrate acceptor was extensively reported [14]. The glycosidic linkage between the sialic acid and the acceptor carbohydrate is extremely controlled by the type of sialyltransferase selected. The transfer of sialic acid residue from CMP-Neu5Ac to 3- and 6-positions of terminal Gal of lactose glycopolymer (**3**), by enzymes α2,3-sialyltransferase and α2,6-sialyltransferase afforded sialyllactose-containing glycopolymer α2,6SGP (**4**) and α2,3SGP (**5**), respectively (Fig. 2).

The resultant SGPs were characterized by ^{1}H NMR spectra. The successful sialylation of lactose glycopolymer was confirmed by the signals of protons from Neu5Ac (1.95 ppm, CH_3-Neu5NAc and 2.60 ppm, H_{3eq}-Neu5Ac), the degree of sialylation and the polymer length as well were calculated also using the ^{1}HMR spectra by comparing the integration value of proton signals from aromatic protons (7.21 and 7.06 ppm), anomeric protons (4.36 and 4.33 ppm) of Gal and Glc, and C3-equatorial proton (2.60 pm) of Neu5Ac as shown in Fig. 3. By comparing the integrated signal from the anomeric protons of Gal and Glc (4.36 and 4.43 ppm,) with that of the C3-equatorial proton of Neu5Ac (2.60 ppm, H3eq-Neu5Ac), the percentages of both sialylations were more than 90 % (Fig. 3).

SGP-based microarrays were fabricated by printing *O*-cyanate chain-end SGP onto amine functionalized glass slide. Next, the specific lectin binding ability of the SGP-based microarray was performed. The resultant α2,3 SGP and α2,6 SGP printed glass slides were incubated with respective FITC-labeled lectins, *Maackia amurensis* (MAA) and *Sambucus nigra* (SNA) (Bioworld) in PBST buffer (pH 7.4) followed by extensive washing with PBST buffer (pH 7.4) and finally the glass slides were subjected to fluorescent imaging, respectively. To confirm the specific binding, competitive inhibition assays were performed by incubating the α2,3 SGP arrayed glass slides with lectin MAA that was pre-incubated with α2,6-sialyllactose, α2,3-sialyllactose and free sialic acid, respectively. Same pattern of inhibition assays were performed by incubating α2,6 SGP arrayed glass slides with SNA that was pre-incubated with α2,6-sialyllactose, α2,3-sialyllactose and free sialic acid, respectively. In addition, in order to determine the sialylation of lactose glycopolymer (**3**), both α2,3 SGP and α2,6 SGP printed glass slides were incubated with lectin PNA that has specificity to α-galactose on the starting glycopolymer (**3**). Overall, these observations (Fig. 4) indicate that arrayed SGPs are capable of distinguishing different glycan-binding protein based on their sialic acid linkage to the galactose in the SGPs.

The interaction of lectin with orientally immobilized SGPs was also investigated with SPR (BI 2000, Biosensing Instrument). The major advantages of SPR assay are that it is a label free assay and monitors the binding in real time. Initially, the gold sensor chip was functionalized with amine by treating the sensor chip with cysteamine in ethanol overnight, the monolayer formation on the sensor chip was confirmed by increase in the resonance units upon modification of the chip. Next, α2,3 SGP was covalently immobilized onto the amine modified sensor chip. Then, a various concentrations of MAA in PBS buffer (pH 7.4) were injected over the α2,3 SGP modified sensor chip to establish the association and

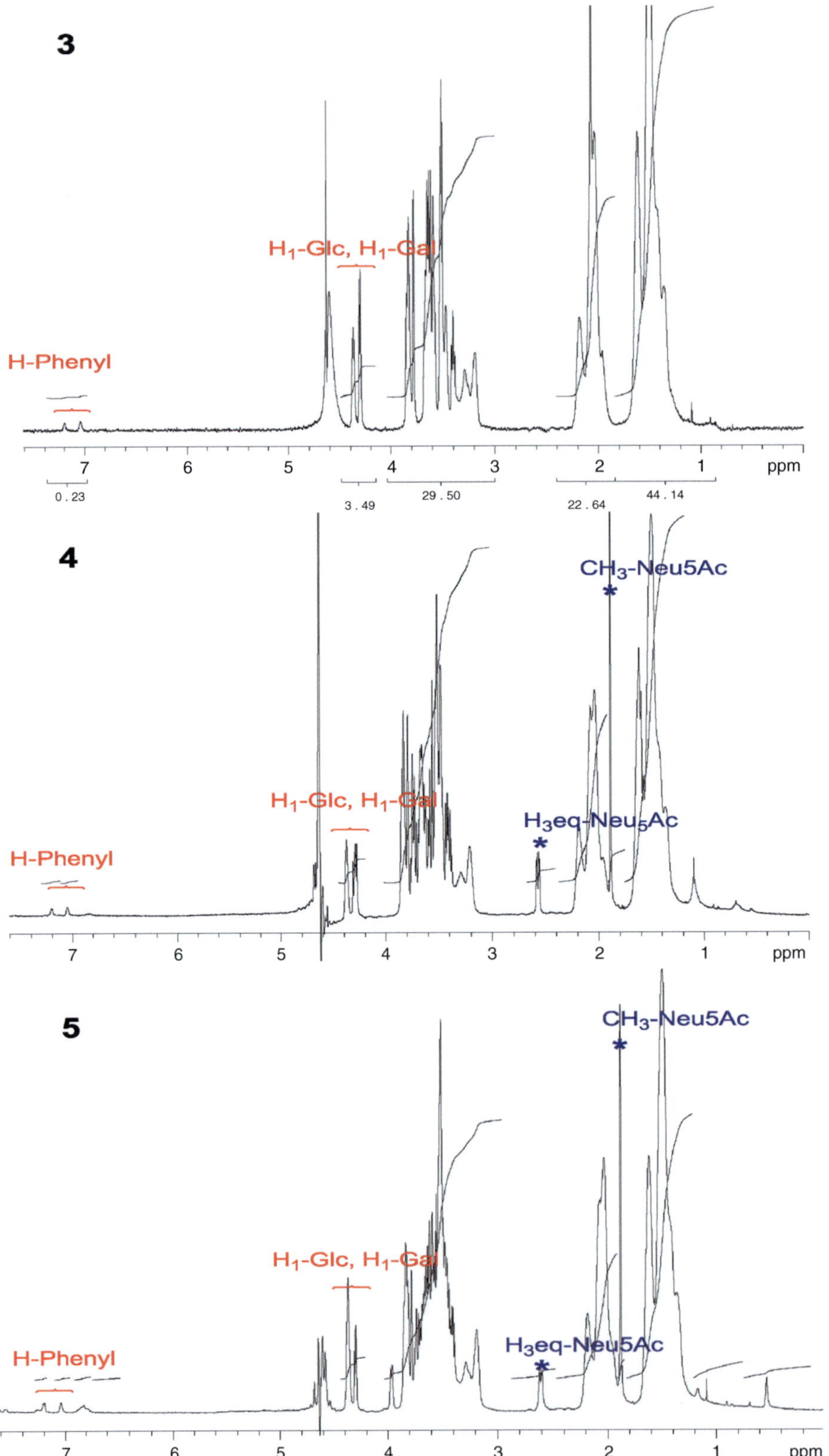

Fig. 3 ^{1}H NMR spectra of glycopolymers in D_2O: (**a**) lactose glycopolymer (**3**), (**b**) α2,6-sialyllactose glycopolymer (**4**), (**c**) α2,3-sialyllactose glycopolymer (**5**)

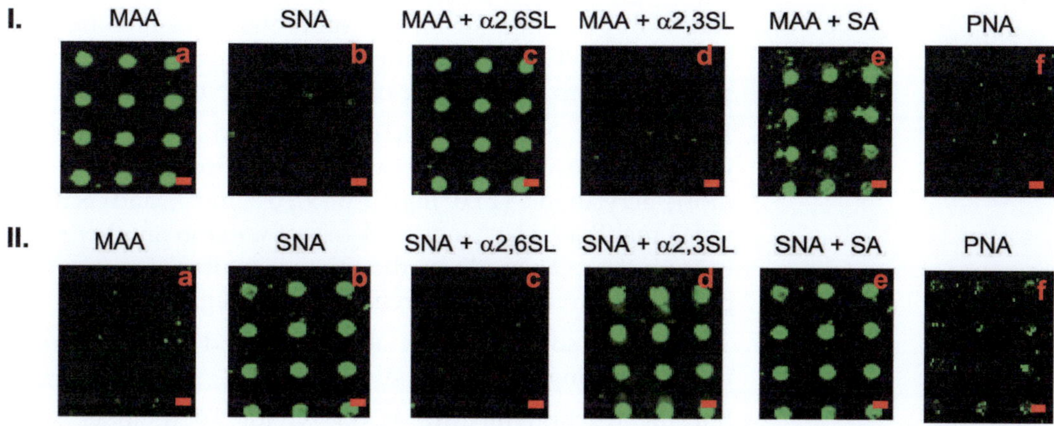

Fig. 4 Fluorescence images of α2,3 SGP and α2,6 SGP microarrays: (**I**) α2,3 SGP array incubated with FITC-labeled lectin, (**a**) MAA, (**b**) SNA, (**c**) MAA pre-incubated with α2,6-sialyllactose (α2,6SL), (**d**) MAA pre-incubated with α2,3-sialyllactose (α2,3SL), (**e**) MAA pre-incubated with free sialic acid (SA), and (**f**) PNA; (**II**) α2,6 SGP array incubated with FITC-labeled lectin, (**a**) MAA, (**b**) SNA, (**c**) SNA pre-incubated with α2,6SL, (**d**) SNA pre-incubated with α2,3SL, (**e**) SNA pre-incubated free SA, and (**f**) PNA. Bar size: 500 μm (*MAA Maackia amurensis, SNA Sambucus nigra, PNA Arachis hypogaea*)

Table 1

Binding kinetics of lectins to immobilized SGPs in SPR assays

Glycopolymer	Lectin	k_a (M^{-1} s^{-1})	k_d (s^{-1})	K_D (M)
α2,3 SGP	MAA	4.39×10^3	9.08×10^{-4}	2.06×10^{-7}
α2,6 SGP	SNA	4.69×10^3	1.25×10^{-2}	2.60×10^{-6}

dissociation constants. Same experiment was conducted using α26 SGP with SNA lectin (Table 1).

With the binding specificity confirmed for immobilized SGPs, we next examined their specific binding to influenza viral hemagglutinins (HAs) of avian H_5N_1 (A/Anhui/1/2005(H5N1)) and human H_3N_2 (A/Brisbane/10/2007(H3N2)) with SPR technique, which bind to α2,3 SGP and α2,6 SGP, respectively. Human and avian HA in PBS buffer (pH 7.4) were injected over immobilized α2,3S GP and α2,6 SGP, respectively. As a result, avian HA showed stronger binding (response units) to α2,3 SGP than α2,6 SGP , and avian HA binds to α2,3 SGP much stronger so that there is no dissociation observed from α2,3 SGP.

3.1 Synthesis of Glycopolymer

1. Dissolve 4-chloroaniline (21.6 mg, 1.69×10^{-4} mol), sodium nitrite (14.1 mg, 2.04×10^{-6} mol) in a mixture of 2 mL water and THF (1:1, v/v) in a three-necked flask.
2. Then, degas the mixture under vacuum and refill with argon gas.

3. Add HBF_4 (66 mg, 7.51×10^{-4} mol) and allow it to react for 30 min at 0 °C under Ar atmosphere to generate diazonium salt with red to yellow color observed (*see* **Note 2**).
4. Dissolve 2-*N*-acryloyl-aminoethoxyl 4-*O*-(β-D-galactopyranosyl)-β-D-glucopyranoside (186 mg, 2.65×10^{-5} mol), acrylamide (210 mg, 2.65×10^{-4} mol), and NaOCN (55.2 mg, 8.49×10^{-4} mol) in 1 mL of water in another round bottom flask. Then, degas the mixture by freeze–pump–thaw procedure three times and refill with argon gas.
5. Transfer the solution of the above degassed mixture into the flask containing diazonium salt.
6. Stir the reaction mixture at 65 °C for 16 h under argon atmosphere.
7. Transfer the reaction mixture to a round bottom flask and connect to rotary evaporator to remove the THF.
8. Filter the reaction mixture through syringe filter to remove any precipitates.
9. Dialyze the resultant mixture (3500 da cutoff) against deionized water for 2 days (change the deionized water in a timely manner) at room temperature to remove any inorganic salts and impurities.
10. Lyophilize the solution to afford glycopolymer (248 mg, 60 % converting yield) (*see* **Note 3**).

3.2 Synthesis of Sialyl Glycopolymer

1. Dissolve O-cyanate chain-end functionalized lactose-containing glycopolymer (40 mg, 1 μmol, 10 μmol of lactose) in 2 mL of 0.10 M HEPES buffer (pH 7.4) containing (Triton X-100 (0.5 %), $MgCl_2$ (5 mM), $MnCl_2$ (2 mM), and $ZnCl_2$ (5×10^{-5} M).
2. To the dissolved polymer solution, add CMP-Neu5Ac (6.5 mg, 100 μmol), calf alkaline phosphatase (10 U), bovine serum albumin (0.8 mg), and α2,3-sialyltransferase (0.5 U).
3. Stir the mixture gently at room temperature for 2 days (48 h).
4. Purify the resultant α2,3-sialyllactose glycopolymer (α2,3 SGP) by gel-filtration with Sephadex G-25 column using DI water as an eluent.
5. Confirm the fractions containing SPG with phenol sulfuric acid assay (*see* **Notes 4** and **5**).
6. Follow the same protocol above to synthesize α2,6-sialyllactose glycopolymer (α2,6 SGP) by using α2,6-sialyltransferase.

3.3 Microarray Fabrication of Sialyllactose Glycopolymer on Amine Functionalized Glass Slide

1. Dissolve α2,3 SGP in 0.10 $NaHCO_3$ buffer (pH 10.3) for a concentration of 2.1×10^{-2} mM.
2. Ink MicroCaster microarray tool with α2,3S GP in pH 10.3 $NaHCO_3$ buffer (2.1×10^{-2} mM).
3. Press onto MicroCaster microarray tool amine functionalized glass slides at room temperature for 10 min.
4. Incubate the glass slides in a humidifying chamber at 37 °C for 4 h.
5. Wash the glass slides with 0.10 $NaHCO_3$ buffer (pH 10.3) (30 min, 3 times).
6. Then wash with 0.10 $NaHCO_3$ buffer containing 0.2 % PBST (pH 7.4).
5. Incubate the array glass slides with lectin MAA-FITC (*Maackia amurensis*, 1.4×10^{-3} mM, Bioworld) in 0.10 M PBST buffer (pH 7.4) containing 0.02 % Tween 20 at room temperature for 3 h.
6. Wash the array glass slides with 0.10 M PBST buffer (pH 7.4) containing 0.02 % Tween 20 for 30 min.
7. Finally, subject the glass slides to fluorescent imaging using Typhoon 9410 Variable Model Imager (Amersham Biosciences, USA).
8. As control experiments, print α2,3 SGP on glass slides and incubate with lectin SNA-FITC (*Sambucus nigra*,1.3×10^{-3} mM, Bioworld), PNA-FITC (*Arachis hypogaea*, 1.6×10^{-3} mM, Sigma), MAA-FITC (0.2 mg, 1.4×10^{-3} mM) that is pre-incubated with α2,3-sialyllactose (1.5×10^{-2} mM, V-Labs), α2,6-sialyllactose (1.5×10^{-2} mM, V-Labs), and free sialic acid (3.2×10^{-2} mM, Rose Scientific Ltd) in 0.2 % PBST buffer (pH 7.4) at room temperature for 2 h, respectively, and followed by extensive washing with 0.2 % PBST buffer (pH 7.4) for 30 min and finally subject to fluorescent imaging..
9. The same protocol should be used for α2,6 SGP microarray fabrication and followed by incubation with lectin SNA-FITC (1.3×10^{-3} mM).
10. As control experiments, print α2,6 SGP on glass slides and incubate with lectin MAA-FITC (1.3×10^{-3} mM), lectin SNA-FITC--> (1.4×10^{-3} mM) that is pre-incubated with α2,6-sialyllactose (1.5×10^{-2} mM), α2,3-sialyllactose (1.5×10^{-2} mM), and free sialic acid (3.2×10^{-2} mM), PNA-FITC (*Arachis hypogaea*, 1.6×10^{-3} mM, Sigma) in 0.2 % PBST buffer (pH 7.4) at room temperature for 2 h, respectively, and followed by extensive washing with 0.2 % PBST buffer (pH 7.4) for 30 min and finally subject to fluorescent imaging.

3.4 SPR Analysis of Specific MAA and SNA Binding to Immobilized α2,3 SPG and α2,6 SGP

1. The specific binding of lectin to immobilized SGP is analyzed with SPR with a BI 2000 biosensor system (BI Biosensing Instruments).
2. To prepare the sensing surface, clean the commercial SPR gold chip (BI Biosensing Instruments) by rinsing with piranha solution of sulfuric acid and hydrogen peroxide (1:1, v/v), followed by DI water and ethanol (3 times) and dry under nitrogen.
3. Incubate the dried gold chip with 1 mg/mL of cystamine solution in ethanol at 4 °C overnight and followed by washing with ethanol and DI water to remove weakly adsorbed cystamine to generate amine functionalized gold chip surface.
4. Assemble the amine functionalized gold chip onto the chip holder and dock into the SPR instrument.
5. To covalently immobilize α2,3SGP onto the amine functionalized gold chip surface, inject the polymer solution (2 mg/mL, $NaHCO_3$ buffer (pH 10.3)) at a flow rate of 10 μL/min for 600 s and repeat at least three times until the surface is saturated with the polymer, using $NaHCO_3$ buffer (pH 10.3) as a running buffer.
6. Inject a serial of solution of lectin MAA in concentration ranging from 0.5 to 10 μM in 0.01 mM PBS running buffer (pH 7.4) over the immobilized α2,3SGP at a flow rate of 20 μL/min for 2 min (120 s, association phase).
7. For regeneration of the surface, inject 0.1 M HCl (pH 1.5) at a flow rate of 20 μL/min for about 30 s and repeat twice to remove bound lectin completely if necessary.
8. Determine the association and dissociation constants of the lectin binding to immobilized SGP by standard BI 2000 (scrubber) evaluation software.
9. For the experiments with α2,6 SGP, the same surface modification procedure should be followed as above. For lectin SNA binding, the same procedure should be used as above except different concentrations ranging from 1 to 10 μM must be used.

3.5 Hemagglutinin (HA) Binding to Immobilized SGP

1. Both α2,3 SGP and α2,6 SGP immobilized on the gold sensor chip following the same procedure as Subheading 3.4.
2. Prepare solutions of H5N1 HA (A/Anhui/1/2005(H5N1)) and H3N2 HA (A/Brisbane/10/2007(H3N2)) (Sino Biological, China) in 0.01 mM PBS buffer (pH 7.4, 420 nM),
3. Inject the HA solutions over the two immobilized α2,3S GP and α2,6 SGP at a flow rate of 20 μL/min for 180 s, respectively.

4. As control, both H5N1 HA and H3N2 HA are pre-incubated with α 2,3-sialyllactose (PBS buffer (pH 7.4), 5×10^{-6} mM) and α 2,6-sialyllactose (PBS buffer (pH 7.4), 5×10^{-6} mM) for 30 min, and then inject over immobilized α2,3 SGP and α2,6 SGP, respectively.
5. Inject 100 mM NaOH for regeneration of surface between each HA injections.
6. Determine the association and dissociation constants of the protein to immobilized SGP by standard BI 2000 (scrubber) evaluation software.

4 Notes

1. 2-*N*-acryloyl-aminoethoxyl 4-*O*-(*β*-D-galactopyranosyl)-*β*-D-glucopyranoside) should be synthesized from lactose peracetate as described previously [11]. Briefly, glycosylation of lactose peracetate (Sigma) with 2-azidoethanol in the presence of boron trifluoride etherate gives the *β*-azidoethyl-lactose peracetate, which is then converted to *β*-aminoethyl-lactose via catalytic hydrogenation and deprotection with aqueous sodium hydrogen carbonate. Finally, *N*-acryloylation of an aminoethyl aglycon with acryloyl chloride gives the product.
2. The red to yellow color is the indicator of diazonium salt formation when HBF_4 is added to the solution of 4-chloroaniline, sodium nitrite in a mixture of 2 mL water and THF (1:1, v/v).
3. The resultant glycopolymer must be characterized by ^{1}H NMR spectroscopy. It is noteworthy that the phenyl group at the end of the glycopolymer provides a convenient mechanism for calculating carbohydrate content and the molar mass through comparison of integrated phenyl proton peaks with those of carbohydrate and backbone protons, respectively (Fig. 3).
4. To 0.05 mL of each fraction add 0.45 mL of DI water, 0.5 mL of 5 % aq. phenol solution, and 2.5 mL of H_2SO_4 (96 %). Change in color from colorless to yellow or brown indicates SGP in the fraction.
5. The resultant SGPs can be characterized by ^{1}H NMR spectra. The successful sialylation of lactose glycopolymer is confirmed by the signals of protons from Neu5Ac (1.95 ppm, CH3-Neu5NAc and 2.60 ppm, H_{3eq}-Neu5Ac), the degree of sialylation and the polymer length as well are calculated also using the ^{1}HMR spectra by comparing the integration value of proton signals from aromatic protons (7.21 and 7.06 ppm), anomeric protons (4.36 and 4.33 ppm) of Gal and Glc, and C3-equatorial proton (2.60 pm) of Neu5Ac as shown in Fig. 3. By comparing the integrated signal from the anomeric protons

of Gal and Glc (4.36 and 4.43 ppm) with that of the C3-equatorial proton of Neu5Ac (2.60 ppm, H_{3eq}-Neu5Ac), the percentages of both sialylations are more than 90 %.

Acknowledgements

This work was supported from AHA Grant-in Aid grant, Ohio Research Scholars Program grant, NSF MRI grant, and Doctoral Dissertation Research Expense Award Program (Cleveland State University). Thanks to Dr. Dale Ray at Case NMR Center for NMR studies.

References

1. Skehel JJ, Wiley DC (2000) Receptor binding and membrane fusion in virus entry: the influenza hemagglutinin. Annu Rev Biochem 69:531–569
2. Palese P, Tobita K, Ueda M, Compans RW (1974) Characterization of a temperature-sensitive influenza B virus mutant defective in neuraminidase. Virology 61:397–410
3. Wiley DC, Skehel JJ (1987) The structure and function of the hemagglutinin membrane glycoprotein of influenza virus. Annu Rev Biochem 56:365–394
4. Murti KG, Webster RG (1986) Distribution of hemagglutinin and neuraminidase on influenza virions as revealed by immunoelectron microscopy. Virology 149:36–43
5. Matrosovich M, Klenk H-D (2003) Natural and synthetic sialic acid-containing inhibitors of influenza virus receptor binding. Rev Med Virol 13:85–97
6. Ting SSR, Chen G, Stenzel MH (2010) Synthesis of glycopolymers and their multivalent recognitions with lectins. Polym Chem 1:1392–1412
7. Voit B, Appelhans D (2010) Glycopolymers of various architectures-more than mimicking nature. Macromol Chem Phys 211:727–735
8. Jiang X, Housni A, Gody G, Boullanger P, Charreyre MT, Delair T, Narain R (2010) Synthesis of biotinylated α-D-mannoside or *N*-acetyl β-D-glucosaminoside decorated gold nanoparticles: study of their biomolecular recognition with Con A and WGA lectins. Bioconjugate Chem 21:521–530
9. Narain R, Armes SNP (2003) Synthesis and aqueous solution properties of novel sugar methacrylate-based homopolymers and block copolymers. Biomacromolecules 4:1746–1758
10. Gupta SS, Raja KS, Kaltgrad E, Strable E, Finn MG (2005) Virus-glycopolymer conjugates by copper(I) catalysis of atom transfer radical polymerization and azide-alkyne cycloaddition. Chem Commun 34:4315–4317
11. Sun X-L, Grande D, Baskaran S, Chaikof EL (2002) Glycosaminoglycan mimetic biomaterials. 4. Synthesis of sulfated lactose-based glycopolymers that exhibit anticoagulant activity. Biomacromolecules 3:1065–1070
12. Brown JM, Rogers CJ, Matson JB, Krishnamurthy C, Rawat M, Hsiesh-wilson LC, Lee SG (2010) End-functionalized glycopolymers as mimetics of chondroitin sulfate proteoglycans. Chem Sci 1:322–325
13. Hou S, Sun X-L, Dong C-M, Chaikof EL (2004) Facile synthesis of chain-end functionalized glycopolymers for site-specific bioconjugation. Bioconjugate Chem 15:954–959
14. Muthana S, Yu H, Huang S (2007) Chemoenzymatic synthesis of size-defined polysaccharides by sialyltransferase-catalyzed block transfer of oligosaccharides. J Am Chem Soc 129:11918–11919

Chapter 17

Glycocalyx Remodeling with Glycopolymer-Based Proteoglycan Mimetics

Mia L. Huang, Raymond A.A. Smith, Greg W. Trieger, and Kamil Godula

Abstract

The cellular glycocalyx controls many of the crucial signaling pathways involved in cellular development. Synthetic materials that can mimic the multivalency and three-dimensional architecture of native glycans serve as important tools for deciphering and exploiting the roles of these glycans. Here we describe a chemical approach for the engineering of growth-factor interactions at the surfaces of stem cells using synthetic glycomimetic materials, with an eye towards promoting their commitment towards specific cell lineages with therapeutic potential.

Key words Glycan microarrays, Glycopolymers, Glycosaminoglycans, Proteoglycans, Stem cells, Stem cell differentiation

1 Introduction

The interactions of growth factors with their cellular receptors are mediated by heparan sulfate proteoglycans (HSPGs) [1]. The glycosaminoglycan (GAG) appendages on proteoglycans recruit various growth factors to the cell surface and help organize their interactions with membrane receptors to activate intracellular signaling and gene expression. In mouse embryonic stem cells (mESCs), GAGs composed of alternating units of variously sulfated glucosamine and uronic acid residues orchestrate the formation of complexes between fibroblast growth factors (FGFs) and their receptors (FGFRs) [2]. The sulfation patterns of GAGs are believed to be responsible for binding to various growth factors [3]. Subsequent phosphorylation of the Extracellular Signal-Regulated Kinases 1 and 2 (Erk1/2) and their associated downstream signaling events result in the differentiation of mESCs into neural precursor cells [4]. mESCs lacking exostosin 1 (Ext1), an enzyme responsible for the biosynthesis of HS, fail to form functional ternary FGF-FGFR-HSPG complexes, and are arrested in their embryonic state [5].

Xue-Long Sun (ed.), *Macro-Glycoligands: Methods and Protocols*, Methods in Molecular Biology, vol. 1367, DOI 10.1007/978-1-4939-3130-9_17, © Springer Science+Business Media New York 2016

Herein, we present a strategy for the identification of synthetic GAG-mimetic polymers that bind FGF2 and a cell-surface engineering strategy to influence stem cell specification. Overall, the procedure begins with the synthesis of the glycopolymers from relevant starting materials using RAFT polymerization (Fig. 1) [6]. Polymer synthesis is modular, and the polymers can be designed to bear azido or lipid terminal functional groups. The azido-terminated polymers are useful for the "copper-free click" conjugation onto cyclooctyne-functionalized surfaces [7]. Libraries of such polymers are microarrayed onto glass slides and evaluated for their ability to bind desired proteins or growth factors. On the other hand, the lipid-functionalized polymers are useful towards cell surface re-engineering. The lipid tail allows for passive insertion onto cell surfaces [8]. Following identification of appropriate binders in the microarray, glycopolymers of the identified glycans are made into lipid-terminated polymers and tested in tissue culture for growth factor binding and stem cell differentiation.

2 Materials

2.1 Chemicals

1. Azadibenzocyclooctyne amine.
2. Azide chain transfer agent is synthesized as described in ref. [9].
3. α,α'-Azoisobutyronitrile (AIBN).
4. Chloroform.
5. Deuterium oxide (D_2O).
6. Dichloromethane (DCM).
7. Diethyl ether.
8. Dimethylformamide (DMF).
9. Dioxane (anhydrous).
10. Lipid chain transfer agent is synthesized as described in [10].
11. Methanol.
12. *N*-butylamine.
13. Phenol.
14. Tetrahydrofuran (THF).
15. *tert*-butyl (3-acrylamidopropoxy)methyl carbamate monomer is synthesized as described in [10] (*see* **Note 1**).
16. Trimethylsilyl chloride (TMS-Cl).

2.2 Other Commercial Materials and Reagents

1. Amicon Ultra Centrifugal Filters (10 K MWCO, 0.5 mL capacity, Millipore).
2. Heparin sodium (Acros).

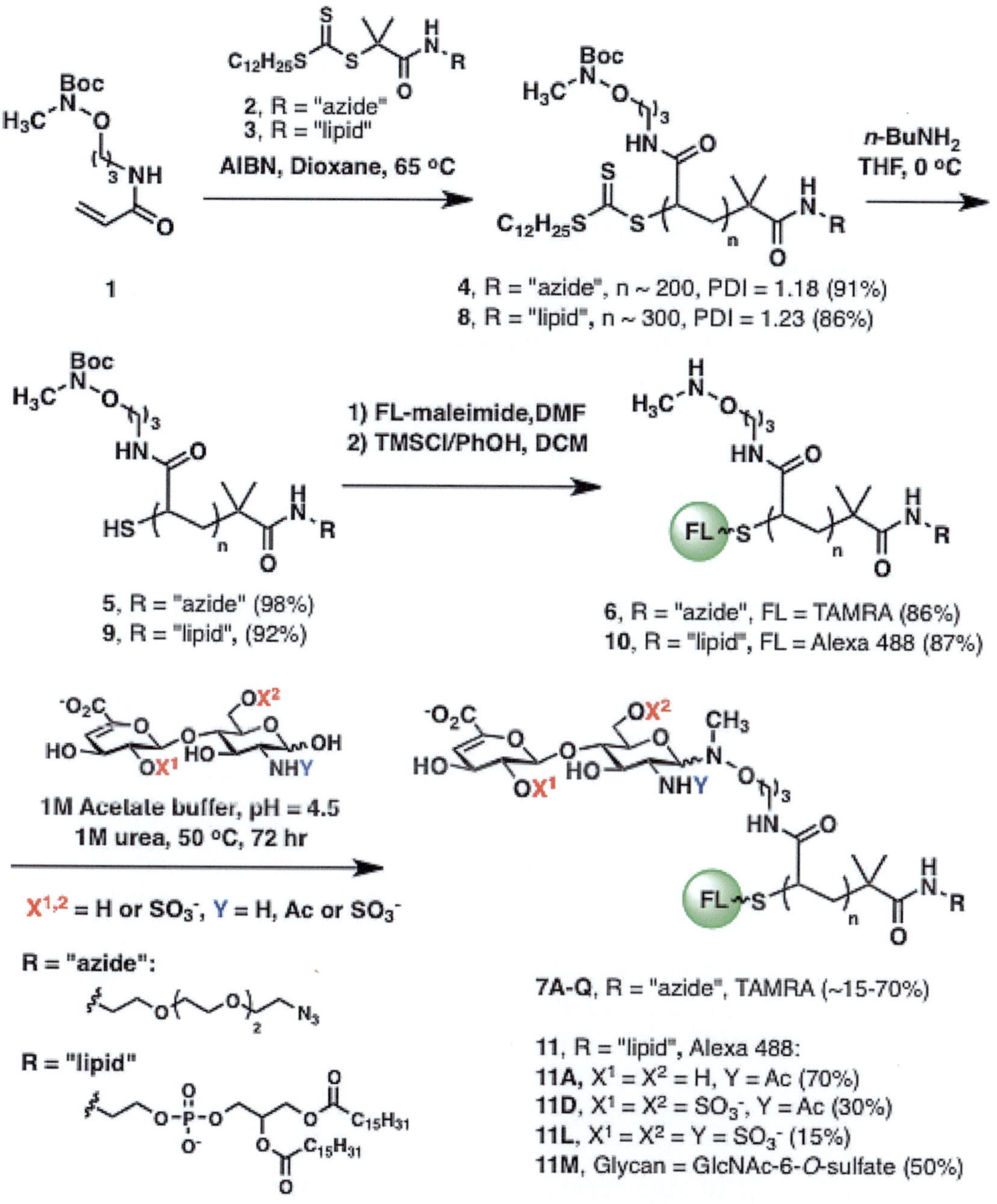

2, R = "azide"
3, R = "lipid"
AIBN, Dioxane, 65 °C

n-BuNH$_2$
THF, 0 °C

1

4, R = "azide", n ~ 200, PDI = 1.18 (91%)
8, R = "lipid", n ~ 300, PDI = 1.23 (86%)

1) FL-maleimide,DMF
2) TMSCl/PhOH, DCM

5, R = "azide" (98%)
9, R = "lipid", (92%)

6, R = "azide", FL = TAMRA (86%)
10, R = "lipid", FL = Alexa 488 (87%)

1M Acetate buffer, pH = 4.5
1M urea, 50 °C, 72 hr

$X^{1,2}$ = H or SO_3^-, Y = H, Ac or SO_3^-

R = "azide":

R = "lipid"

7A-Q, R = "azide", TAMRA (~15-70%)

11, R = "lipid", Alexa 488:
11A, $X^1 = X^2$ = H, Y = Ac (70%)
11D, $X^1 = X^2 = SO_3^-$, Y = Ac (30%)
11L, $X^1 = X^2 = Y = SO_3^-$ (15%)
11M, Glycan = GlcNAc-6-*O*-sulfate (50%)

Fig. 1 Synthesis of fluorescently labeled neoproteoglycans (neoPGs). Polymer backbones can be prepared via RAFT polymerization to bear either azido or lipid end groups. Following trithiocarbonate deprotection, fluorophore conjugation, and Boc sidechain deprotection, the reactive aminooxy groups can be ligated to any glycan with a reducing end. Reprinted with permission from Huang, M.L., Smith, R.A.A., Trieger, G.W., Godula, K. Glycocalyx remodeling with proteoglycan mimetics promotes neural specification in embryonic stem cells. (2014) *Journal of the American Chemical Society* **136**, 10565–10568. Copyright (2014) American Chemical Society

3. PD-10 columns (GE Healthcare).
4. ProLong Gold antifade reagent with DAPI (Cell Signaling Technology).
5. Recombinant Human Fibroblast Growth Factor-Basic (Gibco).
6. Superchip epoxy slides (Thermo Fisher).

2.3 Buffers and Coating Solutions

1. Reaction buffer: 1 M sodium acetate, 1 M urea, pH 4.5.
2. Deuterated PBS buffer: 100 mM sodium phosphate, 150 mM sodium chloride, in D_2O, pD 7.4.
3. PBS buffer: 100 mM sodium phosphate, 2.7 mM potassium chloride, 137 mM NaCl, pH 7.4.
4. Passivating solution: 0.1 % BSA in PBS.
5. Blocking solution: 0.1 % (v/v) Tween in PBS.
6. Slide washing solution: 0.1 % (v/v) Triton X-100 in PBS, pH 7.4.
7. Assay buffer: 1 % BSA, 0.1 % Tween 20, PBS, pH 7.4.
8. ICC buffer: 2 % (v/v) goat serum, 3 % (w/v) BSA, 0.1 % (v/v) Triton X-100 in PBS.

2.4 Antibodies

1. Mouse anti-FGF2 monoclonal antibody, clone bFM-1 (Millipore).
2. Goat anti-mouse IgG, Cy5 (GE Healthcare).
3. Anti-nestin (Millipore cat. #MAB353).
4. Anti-Oct3/4 (Santa Cruz Biotechnology cat. #SC-25401).
5. Anti-mouse Alexa Fluor 555 (Cell Signaling Technology).
6. Anti-rabbit Alexa 488 (Cell Signaling Technology).
7. Anti-phospho-Erk1/2 (Cell Signaling Technology cat. #4370, used at 1:4000 for Western blot).
8. Anti-total Erk1/2 (Cell Signaling Technology cat. #9102 used at 1:6000 for Western blot).
9. Anti-alpha-tubulin (Abcam cat. # ab7750, used at 1:1000 for Western blot).
10. Anti-rabbit HRP (Cell Signaling Technology).
11. Anti-mouse HRP (Cell Signaling Technology).

2.5 Stem Cell Culture and Differentiation

1. Porcine gelatin powder (Sigma Aldrich).
2. Autoclavable glass media bottles.
3. Doubly distilled water.
4. Tissue culture plates (Corning).
5. Mouse embryonic stem cells wild-type E14TG2a and $Ext1^{-/-}$ cell lines [11].

6. Non-enzymatic cell dissociation buffer (Sigma).
7. Trypsin.
8. Maintenance medium: Knockout Dulbecco's Modified Eagle's Medium (KO-DMEM, Invitrogen) supplemented with 10 % (v/v) fetal bovine serum (Gibco), 20 mM L-glutamine (Gibco), 1 % (v/v) nonessential amino acids (Gibco), 55 μM 2-mercaptoethanol (Gibco), and 1000 U/mL leukemia inhibitor factor (LIF, Millipore ESGRO) (*see* **Note 2**).
9. Gelatin stock solution: 1 % (w/v) porcine gelatin in sterile water. Dissolve by autoclaving in glass media bottles (*see* **Note 3**).
10. Neural differentiation medium: 1:1 mixture of neurobasal medium and DMEM/F12 supplemented with 1 % (v/v) B27 (Gibco), 0.5 % (v/v) N2 (Gibco), 55 μM 2-mercaptoethanol, 500 μM L-glutamine, 50 μg/mL BSA, 100 U/mL penicillin, and 100 μg/mL streptomycin [12].

2.6 Western Blot

1. Lysis buffer: 1× cell lysis buffer (Cell Signaling Technology), 1× protease inhibitor cocktail (Cell Signaling Technology), and 1 mM PMSF (Cell Signaling Technology).
2. Laemmli buffer (1×): 2 % (v/v) SDS, 10 % (v/v) glycerol, 5 % (v/v) 2-mercaptoethanol, 0.002 % (w/v) bromophenol blue, 0.0625 M Tris–HCl.
3. Protein ladder (New England Biolabs).
4. Pre-cast gels or self-casted (30 % (19:1) acrylamide–bis-acrylamide, TEMED, APS, Tris, SDS).
5. BCA Protein Quantification Kit (Pierce).
6. 5× running buffer: 72 g glycine, 15 g Tris base in 1 L water.
7. 10× TBS or Tris base solution: 80 g NaCl, 24.2 g of Tris base, 1 L water, pH 7.6.
8. 1× running buffer: 200 mL 5× running buffer, 10 mL 10 % w/v SDS, 790 mL water.
9. 1× transfer buffer: 200 mL 5× Running buffer, 200 mL methanol, 10 mL 10 % SDS, 590 mL water.
10. 1× TBST: 100 mL 10× TBS, 900 mL water, 1 mL Tween 20.
11. Luminata Forte HRP detection reagent (Millipore).
12. ECL hyperfilm (GE).
13. Immobilon-FL 0.45 μM PVDF membrane (Millipore).
14. Restore PLUS Western blot stripping buffer (Thermo Scientific).

3 Methods

3.1 Preparation of Polymers

3.1.1 RAFT Polymerization

1. Prepare apparatus consisting of a flame-dried 10 mL Schlenk flask equipped with a magnetic stirrer, inlet tube provided with a stream of nitrogen, and connected to a vacuum pump.
2. Charge the flame-dried Schlenk flask with the desired (azide or lipid) chain transfer agent (6.2 mg, 0.50 mol% relative to monomer), a solution of AIBN (0.2 g, 0.05 mol% relative to monomer), monomer (570.2 mg), and anhydrous dioxane (443.3 mg) (*see* **Note 4**).
3. Equip the flask with a rubber septum and fill with a stream of nitrogen.
4. Degas the yellow solution with three freeze-pump-thaw cycles.
5. Allow the flask to warm to room temperature, then immerse in an oil bath preheated to 65 °C. Let the solution stir for 12 h (*see* **Note 5**).
6. Dilute the reaction with ether and precipitate with excess hexanes by vigorous stirring. Repeat three times after isolation of the solid residue (by simple decantation) at each step.
7. Remove residual hexanes by successive dissolution and evaporation in chloroform for three times. Dry under high vacuum overnight at room temperature (*see* **Note 6**).

3.1.2 Analysis by GPC

1. GPC is performed on a Hitachi Chromaster system equipped with an RI detector and a 5 μm, mixed bed, 7.8 mm I.D. × 30 cm TSKgel column (Tosoh Bioscience).
2. Polymer is analyzed using an isocratic method with a flow rate of 0.7 mL/min in DMF (0.2 % LiBr, 70 °C) (*see* **Note 7**).
3. Based on the polymer elution profiles, molecular weight distribution differential curves are analyzed using standard approaches to enable the calculation of the polymer molecular weight characteristics (M_n and DI = M_w/M_n).

3.1.3 Fluorophore Conjugation to Polymer Backbones

1. Charge a Schlenk flask equipped with a stir bar with the polymer backbone. Add a 100 mg/mL equivalent of a degassed solution of *n*-butylamine in THF (20 mM).
2. Submerge the Schlenk flask in an ice bath and allow the solution to react for 2 h.
3. Dilute the reaction mixture in ether and precipitate into excess hexanes with vigorous stirring. Perform this step three times, isolating the solid residue at each step.
4. Concentrate the end-deprotected polymer from chloroform three times to remove excess residual hexanes. Dry under vacuum overnight (*see* **Note 8**).

5. Dissolve the polymer (6.35 mg) in a solution of the desired fluorophore–maleimide conjugate in DMF (e.g., TAMRA-maleimide, 2 mM, 1.3 eq, 81.2 μL).
6. Degas the solution with three freeze-pump thaw cycles, and let the reaction stir overnight at room temperature in the dark.
7. Dilute the reaction mixture in ether and precipitate in excess hexanes. Repeat three times, concentrate from chloroform as before, and dry under high vacuum.
8. Store the polymer residue in the freezer until ready for the next step (*see* **Note 9**).

3.1.4 Preparation of Synthetic Glycopolymers

1. To the labeled polymer backbone, add 0.5 mL of freshly prepared solution of TMS-Cl (1 M) and phenol (3 M) in anhydrous DCM. Stir 2 h, in the dark, at room temperature.
2. Add ether (20 mL) to precipitate the Boc-deprotected polymer, and wash three times with fresh ether by centrifugation at 1000 × *g* for 3 min and decantation (*see* **Note 10**).
3. After removing the last ether wash, briefly dry the residue at room temperature with a stream of air or nitrogen.
4. Redissolve the residue in water and pass through a PD-10 column to isolate the polymer from other products.
5. Collect, freeze, and lyophilize the polymer (monitored as the colored band along the column or visible via a UV lamp) (*see* **Note 11**).
6. Redissolve the dried polymer in reaction buffer to form a 200 mM (by side chain) solution. Vortex vigorously and let stand in the dark for a few hours to overnight to dissolve (*see* **Note 12**).
7. Weigh desired amount of glycan (e.g., 1 mg) into a PCR tube. Add the appropriate amount of polymer solution to the PCR tube, such that there is 1.1 eq of glycan per 1 eq side chain. Vortex and quick-spin the mixture to dissolve (*see* **Note 13**).
8. Let the mixture react in a PCR thermocycler for 72 h at 50 °C.
9. Transfer the reaction mixture to an Amicon Ultra Centrifugal Filter, and spin dialyze four times at 6000 × *g*, for 10 min using deuterated phosphate buffer. Discard the flow through and refill the inner tube to 0.5 mL during each wash (*see* **Note 14**).
10. Collect the final solution from the inner filter by inverting into a new conical tube and centrifugation at low speed.
11. Resuspend the sample to a final volume of 0.5–0.6 mL and characterize by ^{1}H NMR and UV analysis to determine ligation efficiencies and polymer concentrations, respectively (*see* **Note 15**).

3.2 Microarray Fabrication

1. In a Coplin jar, rinse the epoxy glass slides with doubly distilled water and spin-dry in a centrifuge (500 rpm, 5 min).
2. Prepare a solution of dibenzyclooctyne-amine (1 mM) and DIPEA (10 mM) in anhydrous DMF. Prepare enough to completely submerge the glass slide in a Coplin jar or petri dish.
3. Let the slides incubate in this solution overnight at room temperature, with a gentle shaking or rocking motion.
4. Remove slides, sonicate in methanol twice for 15 min. Rinse with water and spin-dry. Store at 4 °C until next step.
5. Passivate slides in passivation buffer for 30 min at RT prior to microarray printing.
6. Rinse slides with water and spin-dry (centrifugation at low speed, e.g., 300 rpm, 5 min).
7. Print the microarrayed azido-terminated glycopolymers (10 μM in 0.005 % Tween/PBS, pH 7.4) in replicate spots at 80–85 % relative humidity using a robotic spotter. Scan slide at the appropriate wavelength to inspect spot morphology (*see* **Note 16**).
8. Let slides react overnight at 4 °C in the dark.
9. Wash excess polymers off the slide by vigorously plunging in Slide Washing Solution for 2 min, followed by rocking for 15 min, at room temperature.
10. Wash the slide twice in PBS for 10 min with rocking, rinse with water, and spin-dry.
11. Scan the slide to determine polymer grafting efficiencies (*see* **Note 17**).
12. Fluorescence intensity at 535 nm was measured via a Molecular Devices GenePix 4000B microarray scanner.

3.3 Microarray Evaluation

1. Draw hydrophobic boundaries around each sub-array using the PAP pen, being careful not to draw over the printed spots.
2. To eliminate nonspecific binding, briefly block the wells with assay buffer for 30 min at room temperature.
3. Add a solution of growth factor FGF2 (0.1–1.0 μM) in assay buffer to the desired wells, and incubate at room temperature for 90 min.
4. Wash the well carefully with assay buffer using the edge of a pipet tip and a corner of the well. Repeat four times.
5. Conduct subsequent incubations with the desired primary (mouse anti-FGF2, 1:1000) and secondary antibodies (Cy5-goat-anti-mouse, 1:1000) in assay buffer (*see* **Note 18**).
6. Following all incubations, wash the slide twice for 15 min in PBS, rinse with water, spin-dry, and scan (*see* **Note 19**) (Fig. 2).

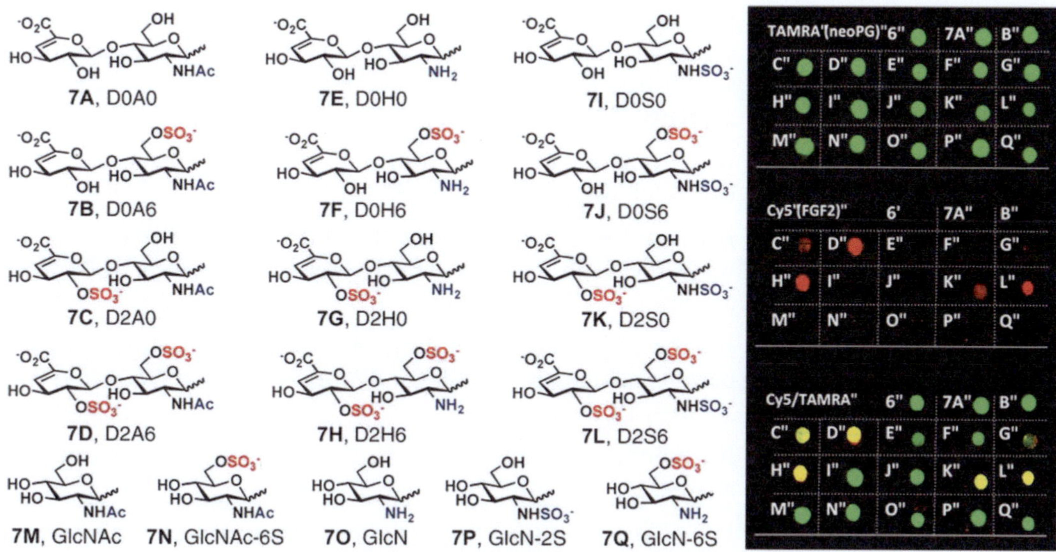

Fig. 2 Microarray analysis of a library of TAMRA-labeled neoproteoglycans (neoPGs) representing most naturally occurring heparan sulfate sulfation motifs identified neoPGs with specificity for FGF2. Reprinted with permission from Huang, M.L., Smith, R.A.A., Trieger, G.W., Godula, K. Glycocalyx remodeling with proteoglycan mimetics promotes neural specification in embryonic stem cells. *Journal of the American Chemical Society* **136**, 10565–10568. Copyright (2014) American Chemical Society

3.4 Tissue Culture

1. Prepare gelatin-coated tissue culture plates by incubating plates or wells with 0.1 % gelatin in PBS solution (diluted 1:10 in sterile PBS or deionized water from gelatin stock solution) for at least 10 min at room temperature. Short-term storage under sterile conditions is also acceptable. Remove gelatin solution prior to use.
2. mESCs are thawed and propagated using maintenance medium in gelatinized sterile tissue-culture-treated well plates.
3. Under normal passaging conditions, cells are split 1:3–1:8 ratios onto new gelatinized well plates, and passaged every 2 days or at 70–80 % confluency.
4. For expansion, cells are split using trypsin or cell dissociation buffer into T-75 or T-175 vented flasks.

3.5 Analysis of neoPG Incorporation

3.5.1 Analysis of neoPG Incorporation by Flow Cytometry

1. A BD FACSCalibur flow cytometer equipped with BD FACSStation is used. Data analysis is conducted using FlowJo (TreeStar). Typical analyses are conducted with 10,000 events in a single run.
2. Detach a (60–80 %) confluent flask of Ext1$^{-/-}$ cells with trypsin and resuspend to 1×10^6 cells/mL in maintenance medium. Seed each well of a 12-well gelatin-coated plate with 1 mL of the cell suspension overnight at 5 % CO_2, 37 °C.

3. Wash each well with PBS (1 mL) and incubate with the desired solution of lipid-terminated neoPG (0.5 mL) resuspended in serum-free KO-DMEM for desired periods of time (e.g., 1 h. *see* **Note 20**).
4. Wash cells 2× with PBS and detach the cells from each well using cell dissociation buffer (1 mL).
5. Harvest the cells from each well into sterile microtubes, and wash the cells 2× with PBS by centrifugation at 300 × *g* for 3–5 min.
6. Fix the cells by resuspending the cell pellet in 300 μL of 4 % paraformaldehyde in PBS for 30–60 min on ice. Wash the cells 2× with PBS.
7. Resuspend the cells in 0.1 % BSA/PBS for flow cytometry analysis.
8. For each condition, conduct at least two replicates, and repeat the assay at least twice (Fig. 3a).

3.5.2 Analysis of neoPG Cell Surface Retention

1. Seed cells and incubate with neoPG and controls as before in a twelve-well plate (*see* Subheading 3.5, **steps 2** and **3**).
2. Wash cells 2× with PBS, then incubate with fresh complete maintenance medium for desired periods of time (e.g., 0, 2, 8, and 17 h) at 37 °C, 5 % CO_2.
3. Following this incubation, repeat steps for harvesting and fixing cells (*see* Subheading 3.5.1, **steps 4–8**) (Fig. 3b).

3.6 Analysis of FGF2 Binding Via Immunocytochemistry

1. Seed 10,000 cells/cm^2 of a gelatin-coated 24-well plate. and incubate with neoPG and controls as before in a 24-well plate (*see* Subheading 3.5, **steps 2** and **3**).
2. Incubate with lipid-terminated neoPG solution in KO-DMEM for desired period of time.
3. Remove media and wash cells 2× with PBS. Fix with 4 % paraformaldehyde in PBS solution for 1 h at 4 °C.
4. Wash cells 2× PBS, block with 2% BSA/PBS for 1 h at 4 °C.
5. Incubate cells with FGF2 (450 nM in 2% BSA/PBS) for 1 h at 4 °C.
6. Wash cells 2× with PBS. Immunostain with rabbit anti-FGF2 (1:1000) and corresponding secondary antibody (1:1000, Alexa Fluor 555 anti-rabbit antibody) for 1 h (each incubation) at 4 °C, in the dark.
7. After final wash with PBS, wash once with water, and add a drop of anti-fade mounting medium with DAPI. Incubate for 3 h at room temperature or store in the dark at 4 °C until ready for analysis.
8. Image under a fluorescence microscope with relevant filter sets (Fig. 4).

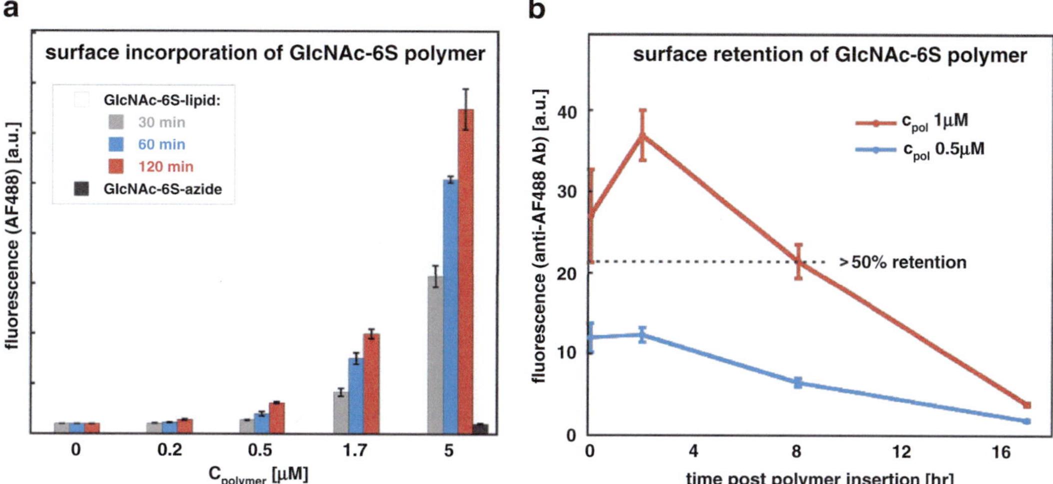

Fig. 3 Flow cytometry analysis of neoPG (**a**) cell surface incorporation and (**b**) cell surface retention. A dose-dependent increase in fluorescence is observed with increasing concentrations of a lipid-terminated polymer (GlcNAc6S-lipid), whereas azide-terminated polymers (GlcNAc-6S-azide) displayed minimal fluorescence. Greater than 50% of fluorescence is retained after 8 h following glycocalyx remodeling. Reprinted with permission from Huang, M.L., Smith, R.A.A., Trieger, G.W., Godula, K. Glycocalyx remodeling with proteoglycan mimetics promotes neural specification in embryonic stem cells. *Journal of the American Chemical Society* **136**, 10565–10568. Copyright (2014) American Chemical Society

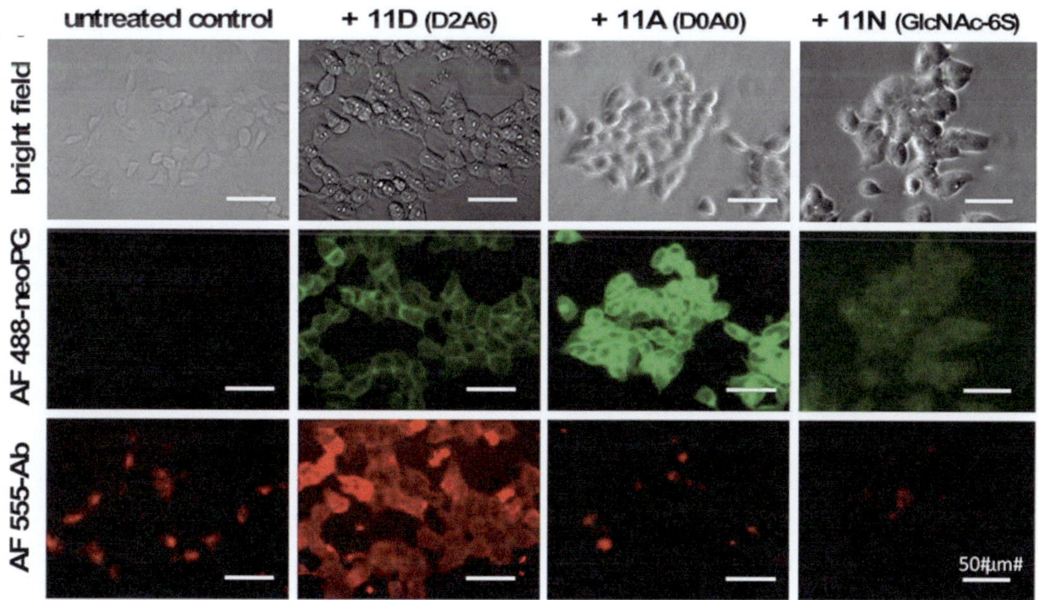

Fig. 4 Fluorescence microscopy analysis of FGF2 binding to cells remodeled with various neoPGs. Cells treated with AF488-labeled neoPGs display green fluorescence. Cells remodeled with FGF2-binding neoPGs (**11D**) show significant fluorescence compared to non-binding neoPGs (**11A** and **11N**) upon treatment with exogenous FGF2 and immunostaining with anti-FGF2 and AF555-labeled secondary antibody. Adapted with permission from Huang, M.L., Smith, R.A.A., Trieger, G.W., Godula, K. Glycocalyx remodeling with proteoglycan mimetics promotes neural specification in embryonic stem cells. *Journal of the American Chemical* Society (2014), **136**, 10565–10568. Copyright (2014) American Chemical Society

3.7 Neural Differentiation of mESCs

1. Harvest cells at 70–80 % confluence.
2. Count and seed cells overnight at 8000 cells/cm^2 in mESC medium onto gelatin-coated tissue culture-treated well plates (for microscopic evaluation, use 24-well plates).
3. Remove medium the following day and replace with neural differentiation medium with or without neoPGs, washing the monolayer twice with PBS prior to media replacement (day 0). Incubate with neoPGs for desired periods of time (e.g., 1 h) at 37 °C, 5 % CO_2, and wash cells three times with PBS. Replace PBS with neural differentiation medium (*see* **Note 21**).
4. Monitor cells daily. Replace with fresh neural differentiation medium whenever necessary (e.g., when media turns yellow) by first washing cells once with PBS (*see* **Note 22**).
5. Harvest cells for microscopy analysis (generally at day 6 or onwards) by fixation with 4 % paraformaldehyde for 10 min at RT, then blocking/permeabilizing with ICC buffer for 1 h at room temperature.
6. Immunostain for embryonic and neural differentiation markers. Perform primary (anti-Oct3/4 at 1:100 dilution and anti-nestin at 1:250 dilution) and secondary antibody incubations in ICC buffer, for 1 h at room temperature (*see* **Note 23**).
7. Follow **steps** 7 and **8** of Subheading 3.6 (Fig. 5).

3.8 FGF2 Stimulation Analysis Via Western Blotting

3.8.1 FGF2 Stimulation

1. Seed cells in gelatin-coated six-well plates at a density of 1×10^5 cells/cm^2 in maintenance medium (*see* **Note 24**).
2. Aspirate medium and replace with maintenance medium with out FBS to serum-starve the cells. Incubate overnight.
3. Aspirate medium and wash cells three times with PBS.
4. Add lipid-neoPGs in KO-DMEM (or KO-DMEM alone as negative control) at desired concentration (e.g., 1 μM). Incubate for 1 h at 37 °C, 5 % CO_2.
5. Remove solutions, wash three times with PBS.
6. Perform stimulation by adding 25 ng/mL of FGF2 in KO-DMEM with or without heparin (5 μg/mL). Incubate for 15 min at 5 % CO_2, 37 °C.
7. Place plate directly on ice. Aspirate medium and wash cells twice with ice-cold PBS. Aspirate PBS.
8. Add 50 μL of ice-cold lysis buffer to each well, and detach cells from plate with a cell scraper.
9. Collect the suspension into an eppendorf, and incubate on ice for 10 min.
10. Centrifuge the suspension at 4 °C, for 10 min at 20,000 xg.
11. Transfer the supernatant to fresh eppendorf. Store at −80 °C if needed.

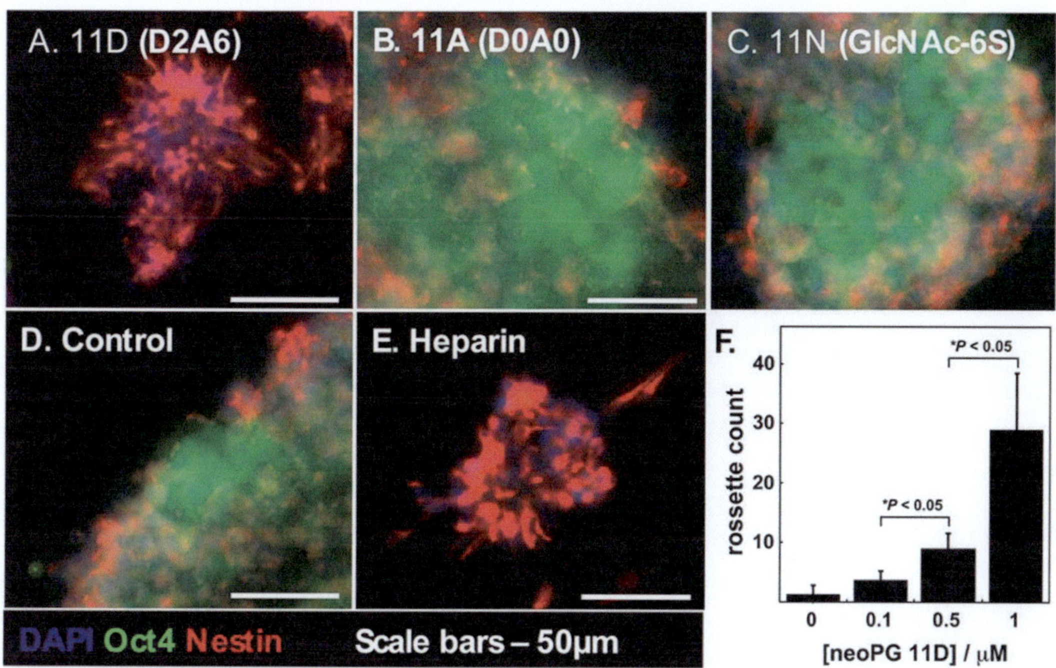

Fig. 5 Neural differentiation of neoPG-remodeled Ext1$^{-/-}$ mouse embryonic stem cells remodeled with neoPGs. (**a**) NeoPG **11D** with D2A6 sulfation pattern, with affinity for FGF2, promoted differentiation into neural rosettes, whereas cells treated with neoPGs **11A** (**b**) and **11N** (**c**), and untreated cells (**d**), retained their embryonic characteristics. (**e**) Soluble heparin was also found to promote neural differentiation into rosettes. (**f**) Increasing dosages of neoPG **11D** increased neural rosette count. Reprinted with permission from Huang, M.L., Smith, R.A.A., Trieger, G.W., Godula, K. Glycocalyx remodeling with proteoglycan mimetics promotes neural specification in embryonic stem cells. *Journal of the American Chemical* Society (2014) **136**, 10565–10568. Copyright (2014) American Chemical Society

3.8.2 Western Blot

1. Quantify protein content using an aliquot of the previous lysate using a standard BCA assay protocol.
2. Add protein sample (5 μg total) to a new eppendorf tube and dilute to 10 μL total volume with doubly distilled water.
3. Add 10 μL of 2× Laemmli buffer to each sample. Briefly vortex at low speed and centrifuge for a few seconds.
4. Transfer eppendorf tubes to a heat block at 100 °C for 5 min to reduce proteins (*see* **Note 25**).
5. Return the boiled samples to ice and centrifuge at low speed upon cooling.
6. Load sample and protein ladder solution (10 μL) in a pre-cast gel (or self-casted 10 % SDS-PAGE) on a running apparatus. Run gel at 100 V in 1× running buffer for 90–120 min.
7. Transfer the proteins from the PAGE gel onto a nitrocellulose membrane using a transfer cassette. Run transfer at constant 350 mA/90 V for 1 h in transfer buffer (*see* **Note 26**).

8. Once complete, block the nitrocellulose membrane in 5 % BSA in 1× TBST for 1 h at room temperature.
9. Remove blocking solution and add primary antibody solution (anti-phospho-Erk1/2, anti-total Erk1/2, anti-alpha-tubulin) in 5 % BSA in 1× TBST. Incubate overnight at 4 °C, overnight (*see* **Note 27**).
10. Wash membranes three times with 1× TBST. Remove last wash.
11. Add HRP-conjugated secondary antibodies (anti-rabbit HRP, anti-mouse HRP) in 5 % BSA in 1× TBST for 1 h at room temperature.
12. Wash membranes three times with 1× TBST. Remove last wash.
13. Develop membrane using HRP detection reagent and developing film (*see* **Note 28**).
14. Perform densitometry using an appropriate software such as Adobe Photoshop or ImageJ (*see* **Note 29**) (Fig. 6).

4 Notes

1. Characterization data for carbamate monomer: ^{1}H NMR (500 MHz, $CDCl_3$) δ (ppm): 7.63 (br s, 1H), 6.28 (dd, 1H, J=17 Hz, J=2 Hz), 6.11 (dd, 1H, J=17 Hz, J=10 Hz), 5.56 (dd, 1H, J=10 Hz, J=2 Hz), 3.91 (t, J=6 Hz, 2H), 3.50 (m, 2H), 3.09 (s, 3H), 1.78 (dd, 2H, J=10 Hz, J=2 Hz), 1.47 (s, 9H). ^{13}C NMR (125 MHz, $CDCl_3$) δ (ppm): 165.6, 156.9, 131.6, 125.5, 81.9, 72.4, 36.9, 28.3, 26.9. Calculated $C_{12}H_{22}N_2O_4$, 258.16, $[M+Na]^+$: 281.15. HRMS m/z found: 281.1470.
2. To prepare maintenance medium, add all components together except 2-mercaptoethanol and LIF, and sterile filter through a 0.22 μm-membrane filtration device. Add the last two components after filtration.
3. It is critical to ensure that glass media bottles are washed extensively with doubly distilled water to remove detergent when used for preparing gelatin stock solutions. It is suggested to use a dedicated glass media bottle for this purpose that does not encounter soap.
4. It is crucial that all components of this reaction are well-dissolved in the solution. Avoid leaving solids on the side of the flask. The ratio of monomer to chain transfer agent can be tuned to achieve the desired degree of polymerization (DP). In our hands, a 300:1 ratio of monomer to lipid chain transfer agent generated polymers of DP~300.

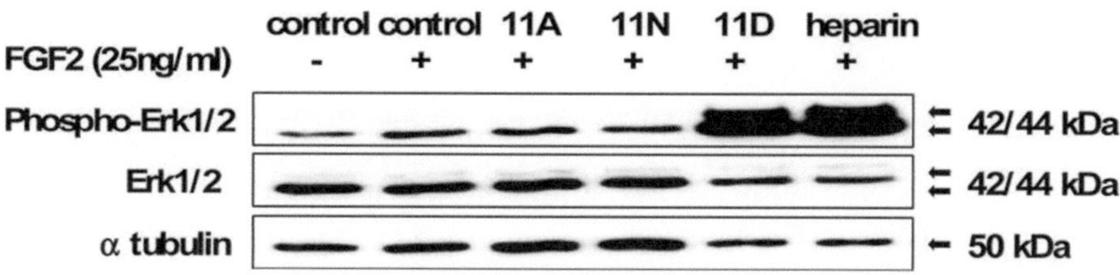

Fig. 6 Glycocalyx remodeling of Ext1$^{-/-}$ mESCs rescued FGF2-mediated signaling. Similar to heparin, cells remodeled with neoPG **11D** significantly promoted Erk phosphorylation compared to controls and neoPGs **11A** and **11N**. Adapted with permission from Huang M.L., Smith, R.A.A., Trieger, G.W., Godula, K. Glycocalyx remodeling with proteoglycan mimetics promotes neural specification in embryonic stem cells. *Journal of the American Chemical* Society (2014), **136**, 10565–10568. Copyright (2014) American Chemical Society

5. The reacted monomer (to form the polymer) is expected to be significantly viscous in form relative to the starting mixture.
6. A pale yellow solid is often isolated in 90% yield. Characterization data for both azide and lipid-terminated polymer products: ^{1}H NMR ($CDCl_3$, 500 MHz) δ (ppm): 3.90–3.65 (bs, 2H), 3.35–2.80 (bm, 5H), 1.80–1.05 (bm, 16H).
7. For the azide-terminated polymer: $M_w = 47.3$ kDa, $M_n = 40.1$ kDa, DI = 1.18, DP ≈ 200. Lipid-terminated polymer: $M_w = 78.5$ kDa, $M_n = 63.9$ kDa, DI = 1.23, DP ≈ 300.
8. A white solid is isolated in ~95 % yield. Successful end-deprotection can be monitored as the reduction of UV absorbance at 310 nm.
9. Determine polymer labeling efficiencies by UV–Vis absorbance at the appropriate fluorophore wavelength. For TAMRA ($\lambda_{max} = 510$ nm), we observed 98% labeling efficiencies.
10. At this stage, the polymer precipitates from ether as a colored fibrous-like residue. Centrifugation helps to clump the residue, such that simple decantation of the solvent is possible.
11. Characterization data for side-chain deprotected polymer. ^{1}H NMR (D_2O, 300 MHz) δ (ppm): 3.94 (bs, 2H), 3.13 (bs, 2H), 2.82 (bs, 3H), 1.75 (6H, bm). It is best to leave the polymer on the lyophilizer right before dissolution in reaction buffer for the next step.
12. A benchtop UV-lamp can be used to monitor the dissolution of clumped polymer residue in reaction buffer. Vortex or shake as needed to dissolve.
13. Glycans used to generate mimics of heparan sulfate were purchased from commercial vendors. These glycans are generated by enzymatic digestion of heparan sulfate. For a description of diGAG nomenclature, see reference [13]. We conduct these reactions in a PCR thermocycler, mainly because of the small

scale of the reaction, and the convenience of weighing small amounts in PCR tubes. Other formats may be acceptable.

14. The centrifugal filters often come with a small amount of glycerol present, which can obscure subsequent NMR analysis. This can be minimized by washing the filter 3× with water (with centrifugation), and leaving the filter to soak overnight with water, and re-centrifugation.
15. Residual acetate peaks from the reaction buffer are often detectable in the NMR analysis of the final product. These are present as a singlet at 1.88 ppm. For diGAGs containing α,β-unsaturated uronic acids, ligation efficiency can be determined by integrating the glycan olefin signal (5.8 ppm) relative to the polymer backbone methyl protons (2.4–2.8 ppm). For monosaccharides, such as GlcNAc-6S, ligation efficiency can be determined by subtracting polymer backbone protons from the total integration in the region 2.5–4.5 ppm, and dividing this difference by the number of glycan protons. Ligation efficiencies from 15 to 70 % have been observed. Highly charged glycans are observed to cause lower ligation efficiencies, due to charge repulsion effects. Some cleavage of the fluorophore from the polymer backbone has been observed (depending on the fluorophore). It is best to re-characterize the polymer at this stage by UV–Vis to determine polymer labeling efficiencies using known weights of the polymer.
16. A standard robotic printer can be used to array spots of the glycopolymers onto the glass slides. We used a non-contact ultrasonic spotter equipped with glass capillary dispensers (GIX Microplotter, Sonoplot, Middleton, WI, USA).
17. Polymer grafting efficiencies can be calculated as the percentage of fluorescence signal retained versus printed onto the slide.
18. Make sure to include negative controls in the experiment to account for background binding of the primary and secondary antibodies. In addition, a control experiment wherein FGF2 is pre-incubated with 1 mg/mL heparin for 15 min can be included.
19. Growth factor binding to each spot/polymer can be calculated as the average ratio of fluorescence intensities at Cy5/TAMRA.
20. The neoPG solution in serum-free KO-DMEM can be sterilized by passage through a low retention volume 0.22 μm sterile syringe filter.
21. Experiments with neoPG are conducted in 1 μM polymer in a 1:1 mixture of neurobasal medium and DMEM/F12. Control experiments with heparin are prepared as a 5 μg/mL solution of heparin in neural differentiation medium.

22. It is recommended that media changes be conducted around the same time each day to minimize cell death. The first media change can be expected to be required around day 2, and daily after then. Cell death is commonly observed at days 3–4.
23. In our hands, the listed antibodies displayed high efficiency, based on the protocol. Other sources of antibodies may also be applicable.
24. We generally allow seeding for 8–10 h for this experiment prior to serum starvation.
25. The lids will tend to pop open during this step. Place a heavy object atop the eppendorf tubes to prevent them from doing so.
26. Gels and membranes must be handled with tweezers at all times. Pre-wet the nitrocellulose/PVDF membranes (appropriately cut to match the dimensions of the gel and cassette) prior to use. Prepare transfer cassette by encasing the nitrocellulose membrane and gel with four sheets of thick filter paper and fiber pads (outer). Add a frozen block and magnetic stirrer to transfer tank to prevent overheating.
27. For sequential antibody staining, strip membranes of primary or secondary antibodies using a Western blot stripping buffer for 15 min at room temperature. Wash membranes three twice with 1× TBST, and block with 5% BSA in TBST for 1 h at room temperature before subsequent antibody staining.
28. Membranes are incubated for 2 min with 1–2 mL of developing reagent. Excess reagent is drained and membranes transferred to a developing cassette. Membranes are developed using ECL Hyperfilm for a period of 10s to 1 min, dependent on signal strength. Films are finally developed using a photodeveloper in a dark room.
29. Normalize both Phospho-Erk1/2 and total Erk1/2 levels to alpha-tubulin levels, then normalize phospho-Erk1/2 levels to total Erk1/2 levels.

References

1. Varki A, Freeze HH, Vacquier VD (2009) Essentials of glycobiology, 2nd edn. CSHL Press, New York, pp. 531–536.
2. Yayon A, Klagsbrun M, Esko JD et al (1991) Cell surface, heparin-like molecules are required for binding of basic fibroblast growth factor to its high affinity receptor. *Cell* **64**:841–848.
3. Kreuger J, Spillman D, Li JP et al (2006) Interactions between heparan sulfate and proteins: the concept of specificity. *J Chem Biol* **174**:323–327.
4. Kunath T, Saba-El-Leil MK, Almousailleakh M et al (2007) FGF stimulation of the Erk1/2 signalling cascade triggers transition of pluripotent embryonic stem cells from self-renewal to lineage commitment. *Development* **134**:2895–2902.
5. Johnson CE, Ward CM, Wilson V et al (2007) Essential alterations of heparan sulfate during the differentiation of embryonic stem cells to Sox1-enhanced green fluorescent protein-expressing neural progenitor cells. *Stem Cells* **25**:1913–1923.

6. Chiefari J, Chong YK, Ercole F et al (1998) Living free-radical polymerization by reversible addition-fragmentation chain transfer: the RAFT process. *Macromolecules* 31:5559–5562
7. Kuzmin A, Poloukhtine A, Wolfert MA et al (2010) Surface functionalization using catalyst-free azide-alkyne cycloaddition. *Bioconjugate Chem* **21**:2076–2085.
8. Rabuka D, Forstner MB, Groves JT et al (2008) Noncovalent cell surface engineering: incorporation of bioactive synthetic glycopolymers into cellular membranes. *J Am Chem Soc* **130**:5947–5953.
9. Huang ML, Smith RAA, Trieger GW et al (2014) Glycocalyx remodeling with proteoglycan mimetics promotes neural specification in embryonic stem cells. *J Am Chem Soc* **136**:10565–10568.
10. Godula K, Umbel ML, Rabuka D et al (2009) Control of the molecular orientation of membrane-anchored biomimetic glycopolymers. *J Am Chem Soc* **131**:10263–10268.
11. Xin L, Wei G, Shi Z et al (2000) Disruption of gastrulation and heparan sulfate biosynthesis in Ext1-deficient mice. *Dev Biol* **224**:299–311.
12. Ying Q-L, Stavridis M, Griffiths D et al (2003) Conversion of embryonic stem cells into neuroectodermal precursors in adherent monoculture. *Nat Biotechnol* **21**:183–186.
13. Lawrence R, Lu H, Rosenberg RD et al (2008) Disaccharide structure code for the easy representation of constituent oligosaccharides from glycosaminoglycans. *Nat Methods* **5**:291–292.

INDEX

A

Acrylamide (AAm) ... 40, 41, 46
Aggrecan ... 70, 71
Amino acid N-carboxyanhydride ... 62, 64, 66
Arachis hypogaea (PAN) ... 40, 197, 201, 203
Atom transfer radical polymerization (ATRP) ... 124, 126, 128, 132, 133, 137, 138, 140, 143–144, 147, 185, 189

B

Biotin ... 5–9, 111, 117–120
Biotinyl-2-aminoethyl methacrylamide hydrochloride (BAEMA) ... 110, 114, 119–121

C

Cationic Ring-opening Polymerization (CROP) ... 49, 50, 52, 53, 55
2-Chloropropyloxyethyl methacrylate ... 138
Chondroitin sulfate ... 74
Click chemistry ... 40, 45, 123–134
Concanavalin A ... 125, 127, 139, 171, 184, 186
4-Cyano-4-(((phenethylthio)carbonothioyl)thio) pentanoic acid (PETTC) ... 90, 93, 94, 100, 101, 105
2-Cyanoprop-2-yl-α-dithionaphthalate (CPDN) ... 150–152
Cyanoxyl-mediated free-radical polymerization ... 3–11

D

Divinylbenzene ... 125
2-(Dodecylthiocarbonothioylthio)-2-methylpropionic acid (DMP) ... 174–175
Dopamine ... 150, 151
Dopamine methacrylamide (DMA) ... 150, 151, 153, 155

E

Emulsifier-free emulsion polymerization ... 138, 140
2-Ethyl-2-oxazoline ... 51

F

Fibroblast growth factors (FGF) ... 208
Flow cytometry ... 217
Fluorescein isothiocyanate (FITC) ... 32, 34, 36, 125, 130, 133, 197, 199, 203

G

Galactose methacrylate (GalSMA) ... 90, 93, 96–100
Gluconamidoethyl methacrylamide hydrochloride (GAEMA) ... 110, 114, 119–121, 158, 159, 161, 164, 165
Glycan array ... 89, 196, 199, 209, 213, 222
N-Glycan, biantennary complex-type sialyloligosaccharide ... 40
Glyco-macroligands ... 196
Glyconanoparticles ... 149, 158
Glycopolymers ... 3, 7–9, 17, 19, 21–23, 34–35, 39, 46–47, 89, 131–133, 137, 150, 151, 158, 184, 186–189, 195, 201–203, 213–214
Glycopolypeptides ... 61–66
Glycosaminoglycan ... 29, 69, 207
Glycosyl azide ... 41, 45
Glycuronan ... 15, 16, 18
Gold nanoparticles ... 158, 160–162, 164–166, 169–178
Graft copolymers ... 69–85
Grubbs' generation 3 catalyst (Grubbs G3) ... 30, 33
Guinea-pig erythrocyte ... 40
Guluronan ... 21

H

HABA-streptavidin assay ... 9
Hemagglutination inhibition (HI) ... 40, 43, 44, 47
Hemagglutinin ... 195, 201, 204
Heparan sulfate ... 69, 207, 215, 222
Heparin ... 74, 210
H_3N_2 ... 40, 44, 197, 201, 204, 205
Hyaluronan/hyaluronic acid ... 74–76, 81
N-(2-Hydroxyethyl)methacrylamide (HEMAm) ... 14–16, 18, 19, 21–25
2-Hydroxypropyl methacrylate (HPMA) ... 90, 93, 95, 96, 101

I

Influenza virus A/Memphis/1/1971 (H3N2) ... 44
Iron oxide ... 149–151, 153, 154
Isourea bond formation ... 4, 196

Xue-Long Sun (ed.), *Macro-Glycoligands: Methods and Protocols*, Methods in Molecular Biology, vol. 1367, DOI 10.1007/978-1-4939-3130-9, © Springer Science+Business Media New York 2016

L

2-Lactobionamidoethyl methacrylamide (LAEMA) 158, 159, 161, 163–166
Lactose 40, 146
Lectin 46, 92, 176, 187–189, 201

M

Macckia amurensis (MAA) 51, 53, 56, 197, 199, 201, 203–204
Mannuronan 15, 20–21
2-(Methacrylamido)glucopyranose (MAG) 150–152

N

Nanoparticles 152–153, 166, 176
N-Carboxyanhydrides (NCAs) 61
Neoproteoglycan 209, 215
Norbornene carboxylic acid derivative 30, 32, 33, 36
Nucleophilic ring-opening polymerization 49–59

O

O-Cyanate 4, 5, 196–199, 202

P

Peanut agglutinin from *Arachis hypogaea* (PNA) 40, 42, 138, 139, 142, 171, 176, 197, 199, 201, 203
Photoiniferter polymerization 138, 141, 144, 145
Poly(2-(methacrylamido)glucopyranose) (PMAG) 150, 152–155
Poly(2-ethyl-2-oxazoline) 49
Poly(2-oxazoline) 49
Polymer brush 126, 128, 129, 185–192
Polymerization-induced self-assembly 89–107
Polypentafluorostyrene 124, 127, 132
N-Propargyl acrylamide 40, 41, 45
Propargyl glycoside 30, 33
Proteoglycans 69–70

Q

Quantum dots 112, 120
Quartz crystal microbalance (QCM) 40, 41, 43, 44, 46

R

Radical polymerization 3–11, 18, 23, 25, 50, 54, 124, 137, 138, 186, 198
Radical thiol-ene photochemistry 62
Reversible Addition-Fragmentation chain Transfer polymerization (RAFT) 14, 15, 18, 19, 21–23, 25, 40, 41, 46, 50, 54, 56–57, 93, 95, 96, 100–101, 110, 114, 119–120, 124, 137, 150, 158, 161, 167, 170, 172–174, 208, 209, 212
Ring-opening metathesis polymerization (ROMP) 29, 30, 32
Ring-opening polymerization 49–59

S

Sambucus nigra (SNA) 197, 199, 201, 203, 204
Sambucus sieboldiana agglutinin (SSA) 40, 42, 44
Sialic acid 195
6'-Sialyllactose (SA-Lac) 39, 40, 42–46
α2,3-Sialyltransferase 197, 198, 202
α2,6-Sialyltransferase 197, 198, 202
Stem cells 207, 209, 211, 215, 217–219, 221
Streptavidin 8–9
Sulfo-galactose 35
Surface Plasmon Resonance (SPR) 162, 177, 184–187, 189–190, 192, 196, 199, 201, 203–204

T

α,ω-Telechelic glycopolymers 4
Telechelic polymers 172
Teoc (2-trimethylsilylethyl carbamate) 29
Thio-β-D-Galactose (GalSH) 90, 97
Thiol-ene 153
Thiol-halogen click chemistry 123–134
Triphenylphosphine 166

V

Versican 71

MIX
Papier aus verantwortungsvollen Quellen
Paper from responsible sources
FSC® C105338

If you have any concerns about our products, you can contact us on
ProductSafety@springernature.com

In case Publisher is established outside the EU, the EU authorized representative is:
Springer Nature Customer Service Center GmbH
Europaplatz 3, 69115 Heidelberg, Germany

Printed by Libri Plureos GmbH
in Hamburg, Germany